HEIDELBERG SCIENCE LIBRARY

Leo Mandelkern

An Introduction to Macromolecules

Second Edition

Springer-Verlag New York Berlin Heidelberg Tokyo

Leo Mandelkern
Professor of Chemistry
The Florida State University
Tallahassee, Florida 32306, U.S.A.

Library of Congress Cataloging in Publication Data
Mandelkern, Leo.
An introduction to macromolecules.
(Heidelberg science library)
Bibliography: p. 156
Includes index.
1. Macromolecules. I. Title. II. Series.
QD381.M37 1983 547.7 83-387

Typeset by Publishers Service, Bozeman, Montana.
Printed and bound by Halliday Lithograph, West Hanover, Massachusetts.
Printed in the United States of America.

9 8 7 6 5 4 3 2 1

ISBN 0-387-90796-3 Springer-Verlag New York Berlin Heidelberg Tokyo
ISBN 3-540-90796-3 Springer-Verlag Berlin Heidelberg New York Tokyo

To Berdie and Our Three Sons

PREFACE TO THE SECOND EDITION

The reception of the original volume by students, pedagogues, and reviewers has been most gratifying. It appears to have both satisfied a need and served a useful educational purpose. Hence, some ten years later it has been deemed advisable to bring it up to date, if only in a slightly expanded form. The purpose for writing this book and its level remain the same.

Many new polymers have been synthesized in the last decade that have found meaningful and novel uses. Examples of these applications are included in this new edition. Major advances have also been made in biophysics and in molecular biology, as well as in our understanding of natural processes on a molecular level. Foremost among these has been the development of recombinant DNA technology. With it has come the potential for large scale synthesis of hormones and proteins. These new developments have also been incorporated into the present volume. It is my hope that this new edition will still have a widespread appeal to students in all of the natural sciences whatever their major interest. It should also be of use and interest to those starting industrial or academic careers who have not had an extensive background in macromolecular science.

Tallahassee, Florida
August, 1982

Leo Mandelkern

PREFACE TO THE FIRST EDITION

The purpose of this volume is to provide beginning students in the natural sciences with a very elementary introduction to the structure and properties of macromolecular substances. Most students, either those who are science majors or those in a liberal arts program, do not usually have the opportunity to learn about this important class of molecules until they take advanced courses at a senior or graduate level. There is thus a serious gap in our educational processes which has become acute with the widespread recognition of the important role played by macromolecules in biological systems, their extensive use in a myriad of articles of commerce, and their intrinsic scientific interest.

We have attempted to rectify this deficiency by presenting a detailed, but non-mathematical, description of all classes of macromolecules. Macromolecules are a unique class of substances in that they are molecules of extremely high molecular weight. When certain common structural features are recognized among the different kinds of macromolecules, they can be treated and analyzed from a unified point of view. With this approach, we can discuss in one volume macromolecules of biological interest as well as synthetic materials such as rubbers, glasses, coatings, and fibers.

It is our hope that this volume will appeal to students in introductory chemistry, biology, and physics courses and will open to them new vistas and horizons that they would not otherwise experience. It is important to emphasize that it is not intended to serve as an introduction to research in this field. If, however, we succeed in stimulating some readers to pursue further study in this area, we shall be extremely gratified and our efforts will have been well rewarded.

The generous assistance and advice from two colleagues is gratefully acknowledged. Professor G. R. Choppin read a portion of the manuscript in draft form and contributed some very expert advice as to the level and content. Professor K. B. Hoffman was kind enough to read the entire manuscript. Her suggestions and criticisms were invaluable. Without them it is doubtful whether the manuscript would have been completed. Finally, it is a pleasure to acknowledge Mrs. Suzanne Knuth and Mrs. Sandra Burkholder, who typed the manuscript and prepared it in final form. The author is indebted to them for their labors and for the skills that they brought to the task.

Tallahassee, Florida
February, 1972

Leo Mandelkern

TABLE OF CONTENTS

Preface to the Second Edition

Preface to the First Edition

1. Introduction 1

2. Structural Features and Preparation 4
   Classification and Definition 4
   Chemical Repeating Units 8
   Copolymers 17
   Preparation 19
   Branching, Cross-Linking, and Network Formation 24

3. Chain Structure 28
   Internal Rotation 28
   Disordered Chain Conformation 34
   Ordered Conformations 40

4. Rubbers and Glasses 54
   Rubbers–Introduction 54
   Structural Basis for Rubber Elasticity 57
   Glass Formation 61
   Chemical Structure and Glass Formation 65
   Plasticizers 66

5. Crystalline Polymers and Fibers 68
   Crystal Structures 68
   Crystallization Process 70
   Undeformed Crystallization–Morphology and Properties 75
   Melting Process–Undeformed Crystallization 82
   Fibers 88
   Mechanochemistry 94

6. Macromolecules of Biological Importance 100
   Introduction 100
   Polypeptides 100
   Conformational Properties of Polypeptides 104
   Proteins–General 109
   Fibrous Proteins 113
   Globular Proteins 121
   Nucleic Acids–General 130
   Deoxyribonucleic Acid–DNA 134
   Ribonucleic Acids–RNA 143

Epilogue: Macromolecules and Man 153

Annotated Bibliography 156

Index 157

An Introduction to Macromolecules

Chapter 1

INTRODUCTION

The history of mankind has been divided into a number of eras on the basis of the material common to the age. Thus, we have passed through the stone, copper, bronze, and iron ages. The modern epoch may very well be looked back on by historians as the Age of Macromolecules. Macromolecules, which are also known as polymers, provide both the necessities and amenities of life. Some macromolecules are found in nature in the form of wood, hide, wool, cotton, silk, and rubber. Others are synthetic, or man-made, and are used in our everyday life in the form of plastics, fibers, rubbers, packaging material, the paper upon which we write, paints, and coatings. Macromolecules such as carbohydrates, proteins, and nucleic acids are fundamental to the biological sustenance of life and are indeed the essence of life itself. Natural occurring polymers have of course been in use for a very long time. It is only recently, however, that we have begun to understand the molecular structure of these materials and the relationship that exists between their structure, properties, and function.

With this new understanding we have been able to use the natural polymers more effectively and have also been able to devise synthetic materials to replace them. Natural occurring systems can be reproduced and we can now even create new macromolecules with properties or combinations of properties that are not paralleled in nature. Thus, we are living in an age in which new materials are constantly replacing older ones. The potential and future benefit to mankind of this undertaking appears to be unlimited. We are entering an era where the concern for our resources and recognition of their limited abundance are of paramount concern. The development of appropriate substitutes for scarce resources is a matter of major importance to the national interest and to the welfare of society. Polymeric substances will undoubtedly play a significant role in the endeavor to develop replacement materials.

A macromolecule (or "giant molecule" or polymer) is a chemical species of very high molecular weight. For example, water, H_2O, a very common low molecular weight substance, has a molecular weight of 18. This means that 1 mole of water, 6.02×10^{23} molecules, will weigh 18 grams. Benzene, a common organic solvent which is another low molecular weight substance, has a molecular weight of 78. However, 1 molecule of natural rubber has a molecular weight of about 1 million, i.e., 10^6. One mole of rubber will therefore weigh 10^6 grams, which is more than 1 ton. This comparison gives us a vivid picture of the difference between polymers and substances of low molecular weight. The word "polymer" comes from the Greek and means many parts. The high molecular weight of a polymer is a result of joining together in long chains, by means of chemical bonds, many thousands of atoms. The same types of covalent chemical bonds that are involved in the simpler low molecular weight organic molecules hold these giant molecules together. The kinds of atoms and their geometric arrangement give each macromolecule its chemical distinctiveness and thus its particular use and function.

The high molecular weight of these substances is also reflected in their molecular size. For example, water has a molecular diameter of 4 Angstrom (Å) units, which corresponds to 4×10^{-8} cm. Similarly, the diameter of the benzene molecule is only about 6 Å. The atoms in the protein hemoglobin, which has a molecular weight of 64,000, are so arranged that the diameter of this molecule is about 65 Å. A synthetic polymer such as polyethylene, having a molecular weight of 100,000, encompasses a very large volume. Its molecular diameter is about 640 Å. The molecular weights of some natural occurring nucleic acids can be in the tens of millions. The molecular sizes of macromolecules are thus very much larger than what we have been accustomed to from our studies of low molecular weight species.

Although polymeric substances have been used by man for many centuries and although the structural principles of organic chemistry were developed in the 1850s, it is only relatively recently that the macromolecular concept has been recognized and accepted. The possibility that covalent structures comprising thousands of atoms could exist as stable well-defined entities was not easily conceived. A great deal of controversy was generated with respect to this concept. Much opposition had to be overcome before there was general acceptance of the macromolecular hypothesis as it is known today. Part of the difficulty was that what we now recognize as polymers could not be readily analyzed by the prevailing classical methods of chemistry. In the second half of the nineteenth century and the early part of the twentieth, the study of polymers was virtually neglected since they could not be easily purified by the accepted methods of distillation and crystallization. They were thought of as a collection of small molecules held together by some peculiar and mysterious intermolecular forces not present in the simpler compounds of low molecular weight. In retrospect, it seems very strange indeed that it was not until the 1920s that serious investigations were undertaken of these substances, which are so essential to man's health, comfort, and well-being.

At about this time physical and physical-chemical methods of analysis were being developed and were achieving a respectability comparable to the more classical methods. Through the Herculean efforts of a few pioneers, notably H. Staudinger in Germany, the macromolecular concept began to emerge and to be gradually, although in some instances grudgingly, accepted. By the mid-1940s the study of macromolecules was well on its way to becoming the quantitative science that it is today. The acceptance of the concept that polymers are true molecular species of very high molecular weight is the primary reason that great advances have been made in the chemistry and physics of these substances in recent years. Because of the common structural features of a large number of atoms covalently linked to one another, it is possible to examine the properties of all kinds of macromolecules from a unified point of view. At the same time one must be careful to recognize the distinguishing chemical features that characterize a specific macromolecule.

In the following chapters we shall derive the benefits of these past endeavors as we examine how the basic principles of chemical bonding and structure manifest themselves as very specific properties and characteristics in macromolecular systems. As we shall see, there are a very large number of diverse properties, functions, and uses that can be realized. These range, for example, from the transmission of genetic information from one generation to another in living organisms to the development of articles of commerce such as rubber-like materials that can function at very high temperatures. The uses of polymers are very widespread and varied. Before we can embark on an analysis of the relations between structure and properties of macromolecular systems, however, we must first classify the major kinds of polymers and describe briefly the ways in which they are formed.

Chapter 2

STRUCTURAL FEATURES AND PREPARATION

Classification and Definition

A polymer chain can be conveniently described by specifying the kind of repeating units present and their spatial arrangement. Molecules composed of chemically and stereochemically identical units (terminal units excepted) are termed homopolymers. When the chain is composed of more than one kind of repeating unit, it is called a copolymer. The chain units can be interconnected in a variety of ways that must, of course, be consistent with the rules of chemical bonding. It is therefore possible for many different geometrical patterns to evolve. The simplest polymer, in terms of chain structure, is the linear homopolymer. As the name implies, the chain units, in this case, are arranged in a linear sequence that can be represented schematically as shown by Figure 2.1. Here *A* represents the repeating or structural

Y---A–A–A–A–A–A–A–A–A–A–A–A–A---X

Figure 2.1. Linear homopolymer.

unit, and *X* and *Y* are the terminal units that may or may not be different from one another and from *A*.

Departures from this simple linear array lead to structures of increasing geometric complexity. A nonlinear or branched structure such as shown in Figure 2.2 can

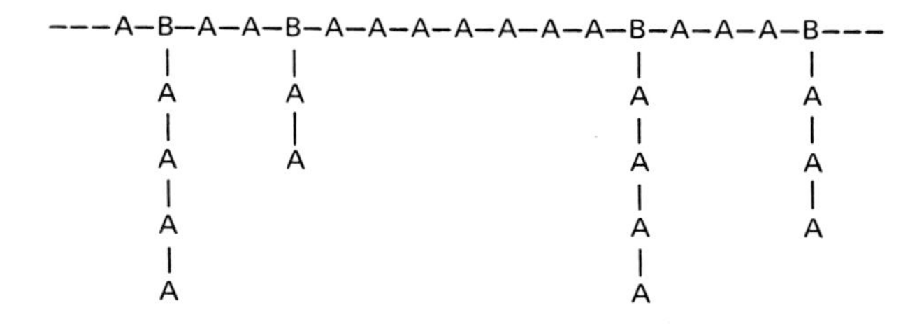

Figure 2.2. Branched polymer.

easily be formed. Here, neither the length of the branches nor their distribution along the main chain has to be uniform. For branching to occur, however, some of the units—the branch points—must possess bonding characteristics that are different from the predominant repeating unit. These differences, however, need not be very large, as is seen in the comparison of the structure of linear and branched polyethylene (Figs. 2.3 and 2.4). Figure 2.4 illustrates what is commonly known as short-chain branching. Here the branches are of small size. Long-chain branches, where the branch lengths are comparable to that of the main chain, or backbone, are also found in many polyethylenes. The chemical differences between linear and branched polyethylene are obviously quite small. Yet, because of these structural differences, these

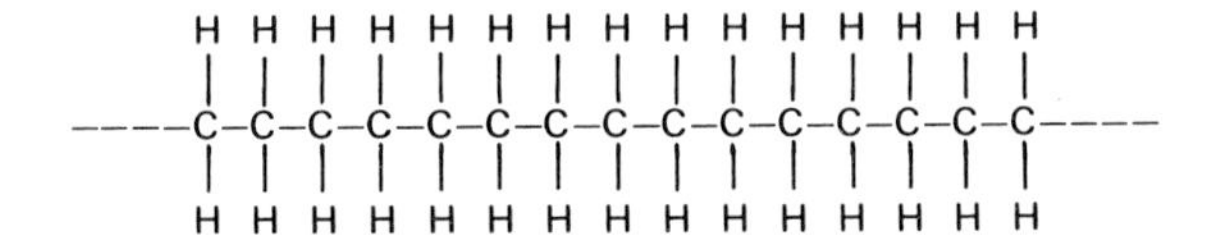

Figure 2.3. Linear polyethylene.

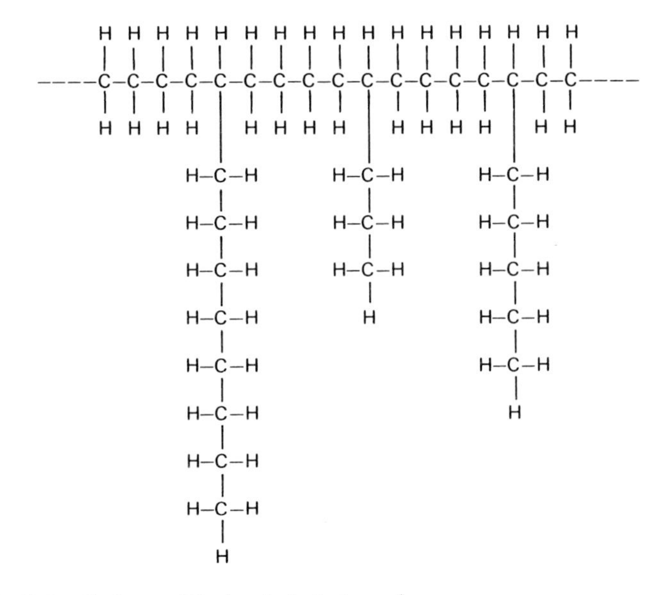

Figure 2.4. Polyethylene with short-chain branches.

two molecules have quite different properties. The relationships between the structure and properties will be discussed in detail in subsequent chapters. Despite their chemical identity, the uses that can be made of these two polymers is, therefore, quite different.

Branched structures are not necessarily restricted to the particular form of Figure 2.2. A more extensive distribution of branched points results in very highly ramified shapes. Thus, a typical structure such as that shown in Figure 2.5 can develop. A continuation of this pattern will lead to a very complicated three-dimensional structure, schematically illustrated in Figure 2.6. Although for branched chains the concept of discrete molecules can still be maintained, it has to be abandoned when considering a three-dimensional or infinite network. A given network encompasses the complete macroscopic domain of the sample. Hence, the expression infinite network.

Nonlinear and network structures can also be prepared from a collection of linear chains by covalently linking together chain units selected from different molecules. Such a system is said to be cross-linked and can be represented as shown in Figure 2.7. Here, X represents the chemical species that covalently links together the A units from different molecular chains. When a sufficient number of units are intermolecularly cross-linked, then an infinite network will be formed.

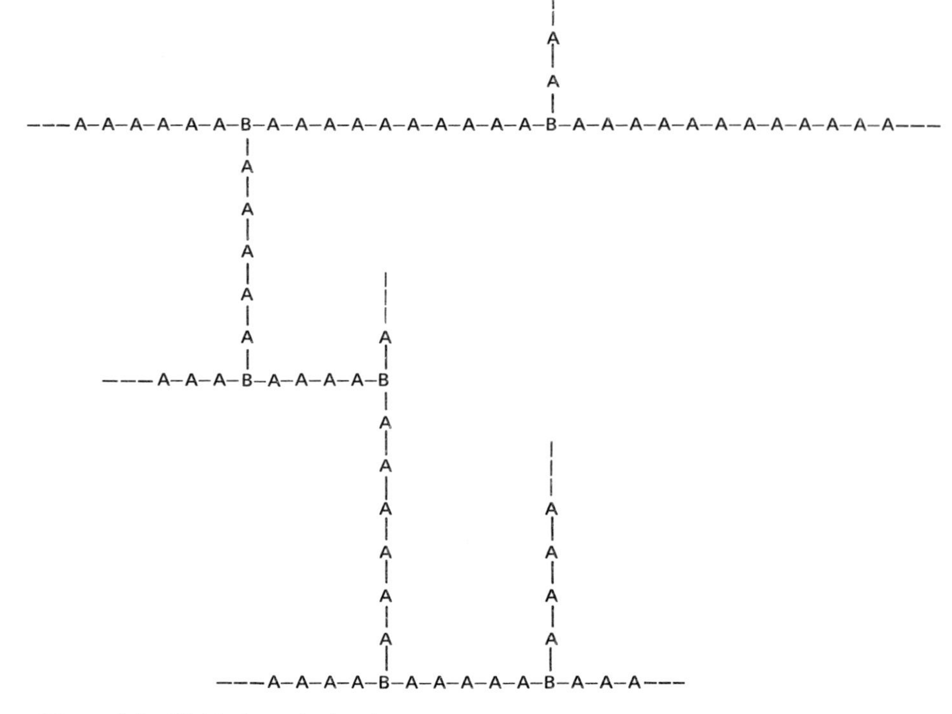

Figure 2.5. Highly branched polymer.

Figure 2.6. Schematic representation of three-dimensional network.

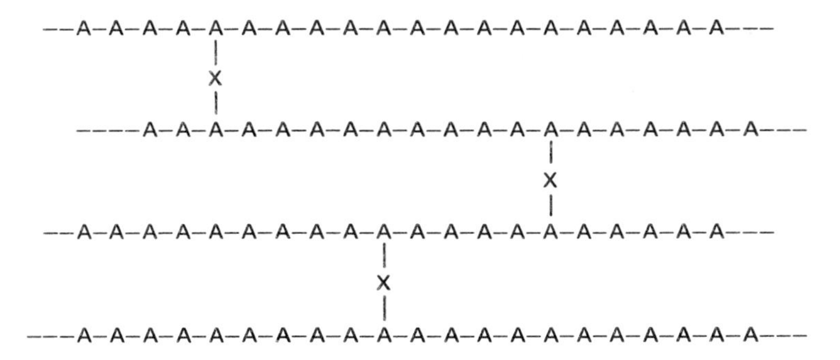

Figure 2.7. Cross-linked system.

The discussion so far has been concentrated on the geometry of structures that can be formed. In some cases a very small fraction of chemically different units has been introduced into the chain. Linear chains can also be made up of different kinds of repeating units. Such chains are called linear copolymers. The properties of a copolymer depend not only on the chemical nature and amounts—or concentration—of the co-units, but also very markedly on how the units are distributed along the chain. For example, given two distinctly different kinds of repeating units forming a chain of fixed composition, there are obviously a large number of different ways in which these units can be arranged in a linear sequence. For simplicity we consider a copolymer to be comprised of A and B type repeating units that differ either chemically or structurally from one another. Three extreme possible arrangements can be classified conveniently. These are an alternating copolymer of A and B such as shown in Figure 2.8. In this case there is a very strong propensity for a given unit to be followed by one of the opposite kind. At the opposite extreme is the block or

---A–B–A–B–A–B–A–B–A–B–A–B–A–B–A–B–A–B–A---

Figure 2.8. Alternating copolymer.

-A–A–A–A–A–A–B–B–B–B–B–B–B–A–A–A–A–B–B–B–B–B–A–A–A–A–A---

Figure 2.9. Ordered copolymer.

ordered copolymer (Fig. 2.9), where there is an overwhelming tendency for a unit to be succeeded by another of the same kind. Here long sequences of one type of unit alternate with long sequences of the other kind. There are several different types of block copolymers, depending upon the number of ordered repeats per molecule. The simplest case would be an A_nB_m type. Here, in each molecule there is one sequence of n A units followed by a sequence of m B units. The case of three ordered sequences per molecule is represented as $A_nB_mA_n$. The most general situation, given in Figure 2.9, can be represented as $(A_nB_m)_x$. In discussing the properties of block copolymers, it is important that the type be specified.

The third major classification is a random copolymer, where the different units are randomly distributed along the chain. Such a linear structure can be represented as shown in Figure 2.10. It is a tribute to the skills of the synthetic polymer chemist

--B–A–A–B–A–A–B–B–A–A–A–B–A–A–B–B–A–A–A–B–A–B–B–A–A–B--

Figure 2.10. Random copolymer.

that these different kinds of chains, with correspondingly different properties, can be produced from the same set of units.

Departing from the restriction of a linear array, branched copolymers, known as "graft polymers," can also be prepared. Such a molecule is schematically illustrated in Figure 2.11. The backbone of the molecule is composed of one type unit, and

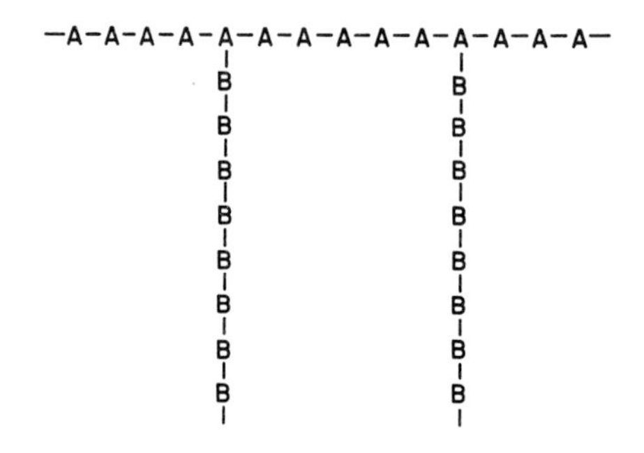

Figure 2.11. Graft copolymer.

the long side chains, or graft, of another. More sophisticated types of graft copolymers have backbones made up of different type repeating units and several distinctly chemically different side groups.

Chemical Repeating Units

A comprehensive listing of macromolecules that have been prepared in the laboratory and found to be useful, as well as those that occur naturally and are involved in biochemical and biological function, would run to many hundreds of entries. Such a compilation would be well beyond our present capacity and interest. Instead, Table 2.1 presents some typical linear homopolymers that were selected to illustrate the tremendous chemical versatility and physical properties that are found in macromolecular systems, which are in turn reflected in their end use or function.

Many of the homopolymers we encounter daily are listed in Table 2-1 along with the chemical formula of the repeating unit, their proper chemical name, and in many cases their common name. Some specialty polymers, which were developed for limited but very important uses, are also included. Strictly speaking, some of the homopolymers listed could be thought of as perfectly alternating copolymers. However, since there is an exact repeat of the structural unit, it is preferable to classify them as homopolymers. Included in this list are naturally occurring polymers

Table 2.1. Homopolymers

Repeating Unit	Poly-	Common or Generic Name
$-\left(\begin{array}{cc} H & H \\ \mid & \mid \\ C- & C- \\ \mid & \mid \\ H & H \end{array}\right)-$	Ethylene	
$-\left(\begin{array}{cc} H & H \\ \mid & \mid \\ C- & C- \\ \mid & \mid \\ H & CH_3 \end{array}\right)-$	Propylene	polyolefin
$-\left(\begin{array}{cc} H & CH_3 \\ \mid & \mid \\ C- & C- \\ \mid & \mid \\ H & C{=}O \\ & \mid \\ & OCH_3 \end{array}\right)-$	Methyl methacrylate	plexiglas, lucite
$-\left(\begin{array}{cc} H & H \\ \mid & \mid \\ C- & C- \\ \mid & \mid \\ H & Cl \end{array}\right)-$	Vinyl chloride	vinyl
$-\left(\begin{array}{cc} H & H \\ \mid & \mid \\ C- & C- \\ \mid & \mid \\ H & C_6H_5 \end{array}\right)-$	Styrene	
$-\left(\begin{array}{cc} H & H \\ \mid & \mid \\ C- & C- \\ \mid & \mid \\ H & CN \end{array}\right)-$	Acrylonitrile	acrylic, orlon
$-\left(\begin{array}{cc} F & F \\ \mid & \mid \\ C- & C- \\ \mid & \mid \\ F & F \end{array}\right)-$	Tetrafluoroethylene	teflon
$-\left(\begin{array}{cc} F & H \\ \mid & \mid \\ C- & C- \\ \mid & \mid \\ F & H \end{array}\right)-$	Vinylidene fluoride	

Table 2.1. *(Continued)*

Repeating Unit	Poly-	Common or Generic Name
$-\!\left(-CH_2-C(CH_3)_2-\right)\!-$	*iso*-Butylene	butyl rubber
$-\!\left(-CH_2-C(CH_3)=CH-CH_2-\right)\!-$	*cis*, 1,4 Isoprene	natural rubber
$-\!\left(-CH_2-C(CH_3)=CH-CH_2-\right)\!-$	*trans*, 1,4 Isoprene	gutta percha
$-\!\left(-CH_2-CCl=CH-CH_2-\right)\!-$	*trans*, 1,4 Chloroprene	neoprene
$-\!\left(-CH_2-CH=CH-CH_2-\right)\!-$	*trans*, 1,4 Butadiene	
$-\!\left(-CH_2-O-\right)\!-$	Methylene oxide	polyether, delrin
$-\!\left(-CH_2-CH_2-O-\right)\!-$	Ethylene oxide	carbowax
$-\!\left(-Si(CH_3)_2-O-\right)\!-$	Dimethyl siloxane	silicone
$-\!\left(-CH_2-CH_2-S(=S)-S(=S)-\right)\!-$	Polysulfide	thiokol rubber
$-\!\left(-C(=O)-(CH_2)_4-C(=O)-NH-(CH_2)_6-NH-\right)\!-$	Hexamethylene adipamide	nylon (6,6)

Table 2.1. *(Continued)*

Repeating Unit	Poly-	Common or Generic Name
$-\!(NH-CO-(CH_2)_5)\!-$	Caprolactam	nylon-6
$-\!(CH_2-CH_2-O-CO-CH_2-CH_2-CH_2-CH_2-CO-O)\!-$	Ethylene adipate	polyester
$-\!(NH-C_6H_4-NH-CO-C_6H_4-CO)\!-$	*p*-Benzamide	aramid, kevlar type
$-\!(NH-C_6H_4-NH-CO-C_6H_4-CO)\!-$ (meta)	*m*-Paramid	aramid, nomex type
$-\!(CH_2-CH_2-O-CO-C_6H_4-CO-O)\!-$	Ethylene terephthalate	dacron, mylar
$-\!(NH-CH_2-CH_2-NH-CO-O-CH_2-CH_2-O)\!-$	Ethylene isocyanate	poly urethane
$-\!(CH=CH-CH=CH)\!-$	Diacetylene	polyacetylene
$-\!(O-C_6H_4-C(CH_3)_2-C_6H_4-O-CO)\!-$	Bisphenol-A-carbonate	poly carbonate, lexan
$-\!(C_6H_2(CH_3)_2-O)\!-$	Dimethyl phenylene oxide	polyphenylene ether

Table 2.1. *(Continued)*

Repeating Unit	Poly-	Common or Generic Name
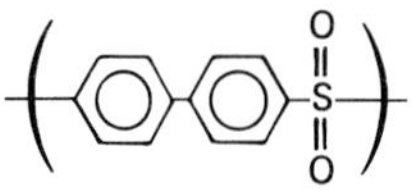	Diphenylether sulfone	polyether sulfone
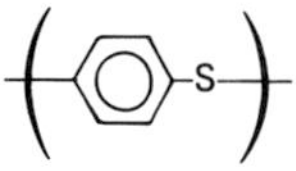	Phenylene sulfide	
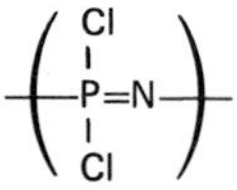 $-(P(Cl)_2=N)-$	Dichlorophosphazene	linear polyphosphazene, inorganic polymer
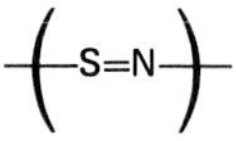 $-(S=N)-$	Sulfur nitride	
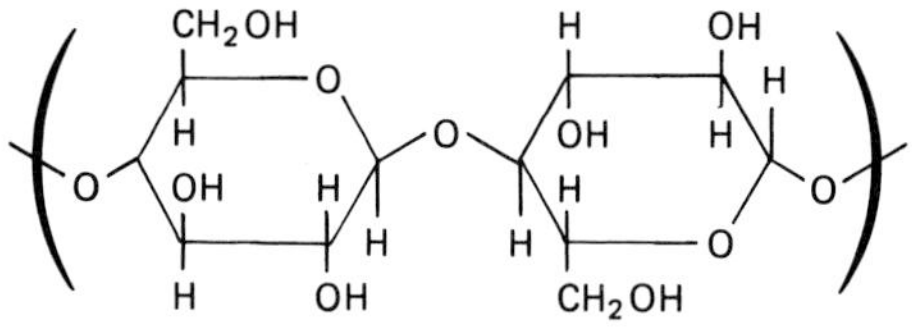	Cellulose	cotton, wood, viscose rayons, cellophane
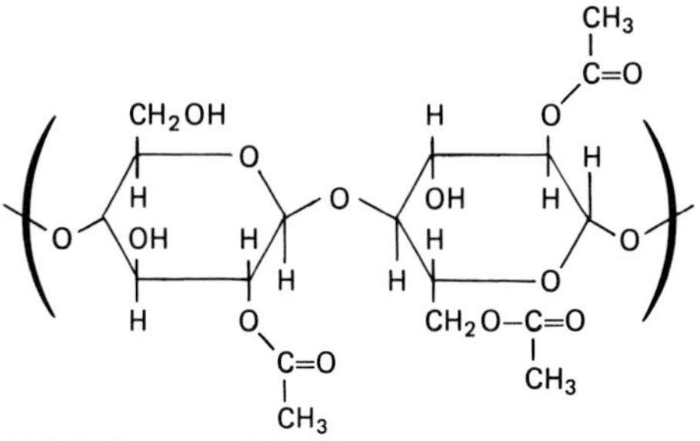	Cellulose acetate[1]	acetate rayon
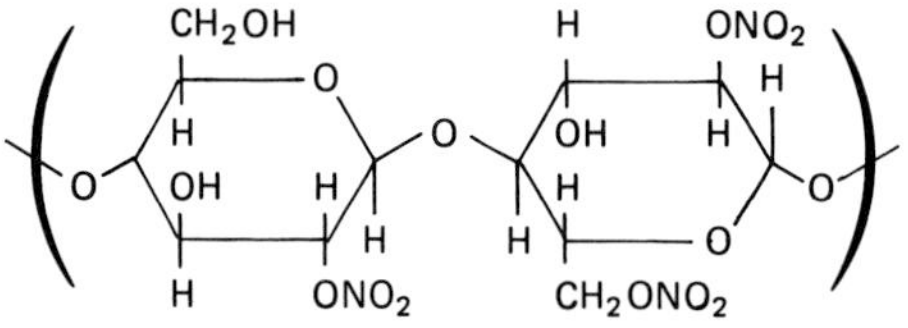	Cellulose nitrate[1]	celluloid, smokeless powder
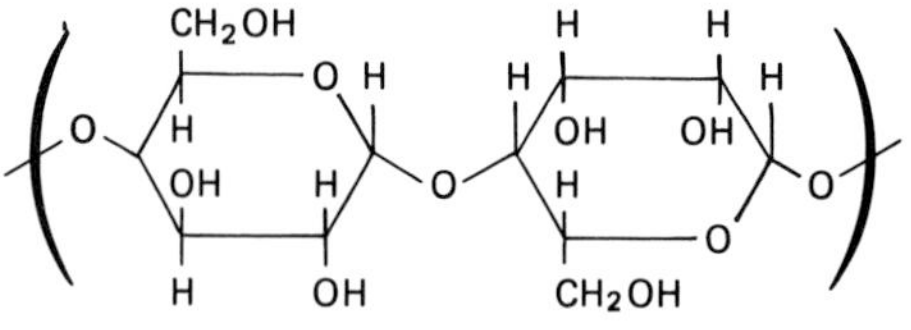	Amylose[2]	starch

Table 2.1. *(Continued)*

Repeating Unit	Poly-	Common or Generic Name
H O H | ‖ | —N—C—C— | CH_3	Poly alanine	polypeptide, poly α-amino acid
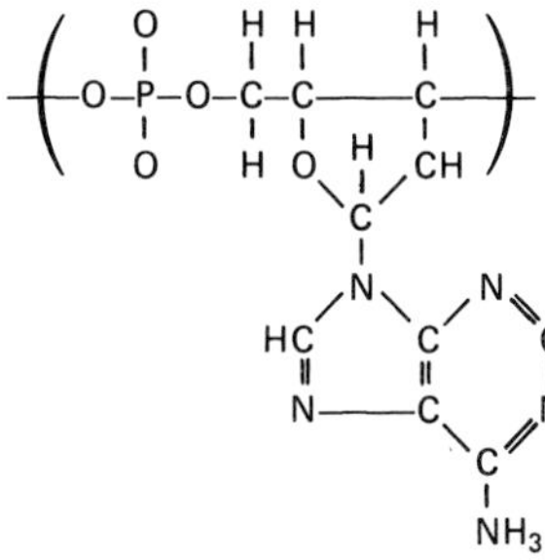 O H H H | | | | —O—P—O—C—C——C— | | | H | O H O | CH C | N N HC C CH ‖ ‖ | N——C N C | NH_3	Poly adenylic acid	nucleic acid

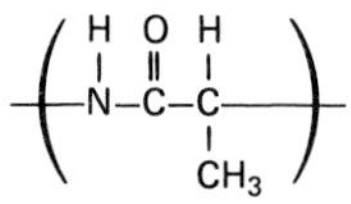

[1] The partially substituted cellulose derivatives are listed. When completely substituted the compounds are called cellulose triacetate and trinitrate respectively.
[2] Chemical studies have shown that there are two kinds of chain structures in starch, each having the same composition. The linear chain is called amylose; the branched chain amylopectin.

such as gutta-percha, natural rubber, and cellulose and its derivatives, which together with the synthetic polymers are widely used in a variety of articles of commerce. Examples of the newly emerging class of useful inorganic polymers are also given. Chain molecules such as starch, which is a polysaccharide, polypeptides, and nucleic acids are also listed. These can be organized into more complex macromolecular structures of great biological and biochemical importance. A more detailed discussion of macromolecular systems of biological importance is presented in Chapter 6. Because of the different kinds of repeating units that can be used and the varied nature of the functional groups that can be attached to the main chain atoms, a great deal of chemical diversity can be built into a polymer chain.

The list of homopolymers in Table 2.1 gives a vivid demonstration of how the chemical nature of the repeating unit influences properties since the polymers have many different end-uses. Before pursuing this point, however, it is instructive to examine some of the differences between the properties of a long-chain molecule and a similarly constituted monomeric substance. The properties of the linear long-chain hydrocarbon, polyethylene, which can be represented by a string of CH_2 units, as in Figure 2.3, serves as a good example. The primitive monomeric analogue is methane,

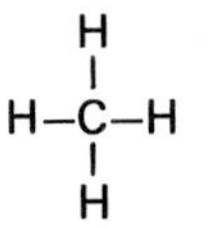

```
   H
   |
H—C—H
   |
   H
```

which is a gas of molecular weight 16. The chain can then be built up by forming successively, ethane,

```
    H H
    | |
  H-C-C-H
    | |
    H H
```

propane,

```
    H H H
    | | |
  H-C-C-C-H
    | | |
    H H H
```

and butane,

```
    H H H H
    | | | |
  H-C-C-C-C-H
    | | | |
    H H H H
```

which also are gases. As the number of carbon atoms in the linear array is increased to six, a very volatile liquid, *n*-hexane, is formed. As the carbon number is further increased, the viscosity of the liquid also increases. When 18 carbon atoms are introduced into the chain, a solid wax-like substance, or normal paraffin, results. The molecular weight of the compound formed at this point is 254. In this way one can build up a long chain of very high molecular weight. Polyethylenes, comprised of hundreds of thousands of chain carbon atoms, and thus having molecular weights in the millions, have been synthesized. Surprising as it may seem, some of the major structural features of the low molecular weight normal paraffins can still be found in these very long chains.

When some of the hydrogen atoms from polyethylene are systematically replaced by other atoms or groups of atoms, the chemical changes that are made also affect the physical properties. The relationships between chemical constitution, structure, and physical properties of long-chain molecules are discussed in more detail in Chapters 4 and 5. As an example, if all the hydrogen atoms are replaced by fluorine atoms, the result is a very interesting polymer, polytetrafluoroethylene, popularly called Teflon. The properties of this polymer are quite different from those of its hydrogen counterpart. It is very inert chemically and resists attack by virtually all chemical reagents. In addition, it is one of the most thermally stable polymers known and does not undergo any physical or chemical changes below temperatures of about 320°C. This polymer possesses a very low coefficient of friction, which makes it very useful as a coating. Teflon also has the property that other materials do not adhere to it, so that it presents a "nonsticky" surface.

Changing the structure of the chain backbone also causes major alterations in properties. As an example, the introduction of aromatic rings into the chain enhances thermal and oxidative stability, raises the softening temperature, and allows for the fabrication of very rigid structural materials. Such polymers are part of a group called engineering plastics. They can replace metals in many applications. They are advantageous because of their light weight, low cost, and ease of fabrication into a variety of objects. Examples are the polycarbonates, polysulfones, polyphenylene ethers, and polyphenylene sulfides. Polyphenylene sulfide, for example, is exceptionally thermally stable. It displays no weight loss up to 500°C and is insoluble in most solvents below 200°C. Its unusual thermal stability and solvent resistance makes it very attractive for many applications.

The aromatic polyamides, or aramid type polymers as they are popularly called, are a relatively recent commercial development. They are finding wide-spread use in high temperature applications. As will be discussed in Chapter 5, they can be formed into fibers of exceptionally high strength with tensile properties superior to the polyesters and aliphatic polyamides. Subtle structural differences, such as meta or para ring substitution as the chain is developed, significantly effects properties.

The large variation in properties that are found in the vinyl polymers comes from the different types of side-groups, such as methyl, phenyl, or chloride that can be introduced into homopolymers, the ease and diversity of copolymerization, and the stereo-irregularity of the chain (see Chapter 3).

Table 2.1 lists two sets of naturally occurring polymers. These are the *cis* and *trans* poly 1,4 isoprenes[1] and the polysaccharides starch[2] and cellulose. In each of these groupings the two species have identical chemical formulas. *Cis*-polyisoprene, commonly called natural rubber, displays the expected elastic properties associated with the name. On the other hand, the *trans* polymer, gutta-percha, is a hard and inelastic substance at room temperature. The only difference between the two chains is the geometrical arrangement about the double bond as is shown in Figure 2.12. For the *cis* polymer the CH_2 groups lie on the same side of the double bond

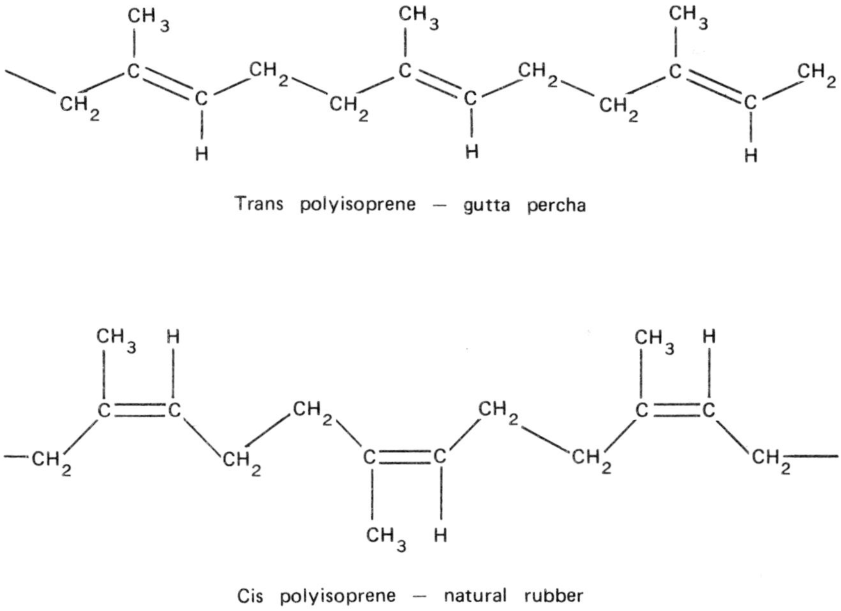

Figure 2.12. Skeletal structure of *cis* and *trans* polyisoprene.

[1] The terminology 1,4 signifies that a linear chain is being propagated through the addition of isoprene units,

$$\underset{1}{CH_2}=\underset{2}{\overset{CH_3}{\overset{|}{C}}}-\underset{3}{CH}=\underset{4}{CH_2}$$

at the 1 and 4 carbon positions.

[2] Although we are using the word "starch," we are in fact referring to the linear amylose chain.

as the chain is perpetuated. For the *trans* polymer the CH_2 groups are arranged on alternate sides of the double bond. Although this difference appears to be only a minor structural alteration, it manifests itself in a set of very different properties.

Similarly, although starch and cellulose chains are made up from the same building block, the glucose unit, they have quite different properties. Starch is readily soluble in many solvents, including water. It is an important foodstuff that is easily digestible by humans, and in the body the polymer is degraded to the monomeric sugar by enzymatic action. On the other hand, cellulose, a major constituent of plants, is a very insoluble substance and is not digestible by man. However, animals such as cows and horses possess the enzymatic apparatus necessary to degrade cellulose to its monomeric unit. The termite plays havoc with our houses by being able to digest and make a meal of wood (cellulose). The reason for these remarkable differences resides again in the geometric nature of the linkage propagating the chain. Figure 2.13 gives a perspective drawing of the glucose unit. The ring is perpendicular

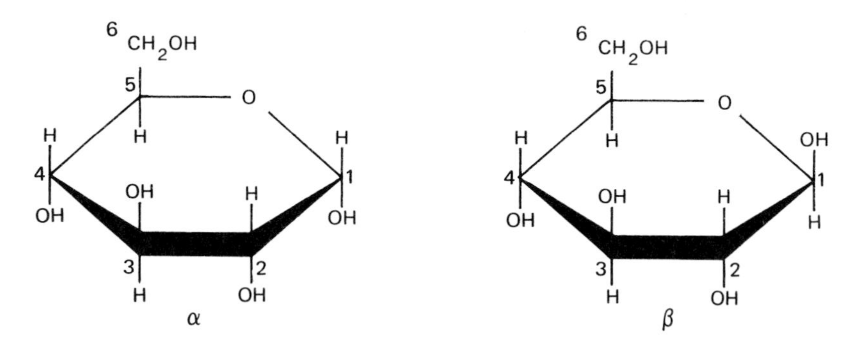

Figure 2.13. Two configurations of the glucose unit.

to the plane of the paper, with the heavily marked bonds pointing outwards. The oxygen atom is then directed to the rear with the ring substituents directed above or below the plane. The two OH groups attached to the carbon atoms in the 1 and 4 positions can either be on the same side of the plane or be opposite to one another. The linkage that perpetuates the chain, the ether type oxygen bond, can be thought of as resulting from the reaction of the 1 and 4 hydroxyl groups from different rings. The relative position of these two groups in the monomers thus determines the important distinguishing feature of the chain, as is shown in Figure 2.14. The

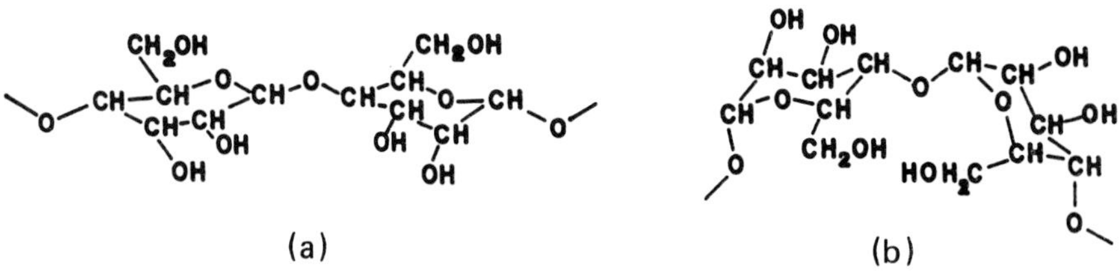

(a) (b)

Figure 2.14. Chain structure of starch and cellulose. (a) Cellulose with β-linkage; (b) Starch with α-linkage.

α-linkage characteristic of starch is always directed below the plane while the β-linkage of cellulose is directed above the plane. This structural difference is, in turn, reflected in the diverse chemical behavior and physical properties of the two chains. In cellulose, the glucose units are so joined as to favor an extended structure. In starch, the bonds are formed so as to impose a bend at each junction between units. Cellulose molecules, being in highly extended forms, can therefore aggregate in parallel array to form stiff fibers of high strength. Such aggregates can thus serve as the structural elements of plants. Starch, on the other hand, is clearly not suited for this purpose. Each of these naturally occurring polymers is always found with its characteristic linkage throughout the chain. A chain with mixed linkages has not as yet been found.

The reaction between the 1 and 6 hydroxyl groups of the monomers, together with the formation of the 1-4 α-linkage in the same molecule, yields the branched constituent of starch, amylopectin.

Copolymers

A further illustration of the chemical virtuosity that can be attained by long-chain molecules manifests itself in copolymers. In Table 2.2 a typical representation of some of the synthetic copolymers is presented.

As we have mentioned, the properties of copolymers depend not only on chemical constitution but also on the sequential distribution of co-units along the chain. For example, with an ordered copolymer desirable properties characteristic of each of the corredponding homopolymers can be incorporated into a single molecular chain. The same properties cannot be obtained, however, merely by mechanically mixing the individual homopolymers. Thus, for example, in the "spandex type" of copolymer the apparently contradictory properties of simultaneously having fiber-like qualities, with the associated high tensile strength, while enjoying the benefits of long-range elasticity are incorporated into one molecular system. Similarly, block copolymers can be prepared that enhance the dyeing capacity of a fiber while still maintaining the other necessary physical properties.

The properties of random and alternating copolymers, on the other hand, are usually quite different from those of the corresponding homopolymers. For example, the random ethylene-propylene copolymer is an elastomer. It has very good resistance to oxidation because of the absence of double bonds, which are found in most other rubber-like polymers. The elastomeric quality of this copolymer is not found in the corresponding homopolymers, polyethylene and polypropylene. These two homopolymers are partly stiff, plastic materials. The butadiene-styrene copolymers, in a 3:1 mole ratio, are also elastomers and are particularly suited for heavy duty automotive use. As the relative amount of styrene is increased, however, the material becomes tough and leathery. In contrast, a block copolymer of styrene and butadiene, although behaving like a vulcanized rubber at room temperature, undergoes plastic flow at high temperatures and can be easily molded into any desired form. The graft copolymer of three species, acrylonitrile, butadiene, and styrene, in which the polybutadiene is grafted on to a random acrylonitrile/styrene copolymer is a very tough, rigid material called ABS. Although many polymeric

Table 2.2. Copolymers

Copolymer	Uses
Butadiene/styrene 3:1 Random	Synthetic rubber; GRS, SBR
Ethylene/propylene 3:1 Random	Elastomer
Ethylene/vinyl acetate	Elvax; wax coatings; drug encapsulate
Ethylene/tetrafluoroethylene Alternating	Wire coating; low flammability
Ethylene/methyl acrylate Random	Elastomer; low temperature, heat and oil resistant
Vinyl chloride/vinyl acetate 8:1 Random	Vinylite; coatings
Vinyl chloride/acrylonitrile 9:1 Random	Dynel; fibers, wigs
Acrylonitrile/vinyl acetate 8.5:1 Random	Acrylic fiber; Orlon, Acrilan
Vinylidene chloride/vinyl chloride Random	Saran: packaging, fibers
Vinylidene fluoride/ hexafluoropropylene 7:3 Random	High temperature rubber
Tetrafluoroethylene/fluoro methyl vinyl ether 1:1 Random	Elastomer; high temperature resistant to most solvents
Polyurethane/polyether Block	Spandex; elastic fiber
Styrene/butadiene Block	Vulcanized rubber at room temperature; undergoes plastic flow at high temperatures; can be molded
Acrylonitrile/butadiene/styrene Graft	ABS; tough plastic, high impact strength

materials are rigid and hard, they are usually also very brittle. Because of the special geometric combination of monomers, this copolymer possesses unique mechanical properties in that it is not only rigid and tough but also has a high impact strength. Hence, it does not fracture very easily. ABS, therefore, has many of the strength properties associated with a metal, but it is also very lightweight and corrosion resistant. It finds a variety of uses, from football helmets to living room furniture. The formation of a graft copolymer is essential in imparting these highly desirable properties. A mixture of polybutadiene with styrene/acrylonitrile copolymer does not possess the high strength.

Acrylic type fibers, which are based on polyacrylonitrile, have other units such as vinyl acetate randomly introduced into the chain. The necessity for a copolymer here is to facilitate the polymer processing and to make the fiber more amenable to dyeing with conventional textile dyes. Vinyl acetate is also a common copolymeric unit introduced into polyvinylchloride. Many fabrication problems are alleviated by this procedure, and the resulting physical properties are improved. The copolymerization of vinyl acetate with ethylene reduces the rigidity of the parent polyethylene homopolymer. Among other uses, this copolymer can serve as a wax additive or as an encapsulting medium for slow-release drug delivery systems.

The random copolymerization of vinylidene fluoride with hexafluoropropylene yields a substance whose mechanical properties are quite different from either of the parent homopolymers. Instead of the rigid, nondeformable material associated with fluorine-containing homopolymers, this copolymer possesses very good elastomeric qualities. At the same time, however, the thermal stability with respect to chain degradation, which is a special characteristic of fluorine-containing polymers, is maintained. That is to say, the integrity of the chain is maintained so that this copolymer behaves as a rubber at very high temperatures. Most of the common elastomers cannot be used at elevated temperatures because of chemical degradation. The copolymers of poly(tetrafluoroethylene) (Teflon) are designed to take advantage of the inherent chemical inertness and thermal stability of the parent homopolymer, but to allow at the same time for greater ease in processing.

The copolymers described here represent but a few from among the hundreds that have been prepared and have found extensive use. These few examples should be sufficient, however, to demonstrate how control of physical and chemical properties can be achieved by appropriate copolymerization when carried out by skilled synthetic chemists.

Preparation

We have been discussing the architectural forms that long-chain molecules can adopt, the great diversity in the chemical types of repeating units that are now available, and in a very general and qualitative manner, the relationship of these factors to properties. We have not as yet considered, however, the processes that enable these giant molecules to be synthesized, either in the laboratory or in nature, from their monomeric components. In discussing polymerization we shall be concerned primarily with the main principles that are involved, and we shall not dwell on any detailed mechanism of a particular reaction.

It is readily apparent and indeed axiomatic that in order to perpetuate a chain to appreciable size, the functionality of the starting monomeric species must be at least two. The meaning of this statement can be seen when we consider two species (a) R–A and (b) R′–B that react with one another to form a third species (c) R–A–B–R′. Thus

$$\underset{\text{(a)}}{\text{R–A}} + \underset{\text{(b)}}{\text{B–R}'} \longrightarrow \underset{\text{(c)}}{\text{R–A–B–R}'} \tag{1}$$

and the new compound (c) is formed by the combination of functional groups. No further reaction of the same type involving species (c) can take place since there are no longer any reactive groups available. Thus, one molecule of species (a) can react once and only once with a molecule of species (b) to give (c). Compounds (a) and (b) are thus said to be monofunctional. A simple example of this scheme, from organic chemistry, is seen in Eq. (2) for the reaction of an acid (1) with an alcohol (2) to give an ester (3) and water,

$$\underset{(1)}{CH_3{-}\overset{\overset{\displaystyle O}{\|}}{C}{-}OH} + \underset{(2)}{HO{-}CH_2{-}CH_3} \longrightarrow \underset{(3)}{CH_3{-}\overset{\overset{\displaystyle O}{\|}}{C}{-}O{-}CH_2CH_3} + H_2O \tag{2}$$

Here, as indicated schematically, the carboxyl group of the acid, $\overset{\overset{\displaystyle O}{\|}}{C}{-}OH$, has reacted with the hydroxyl group, OH of the alcohol to form the ester. In the process, water is given off. In this case both the acid and the alcohol are monofunctional. if either the acid or the alcohol were bifunctional so that, for example, the acid contained two carboxyl groups while the alcohol contained only one hydroxyl group, the reaction scheme indicated by Eq. (2) could proceed one more step. The product of the reaction would, however, still be a low molecular weight species.

Let us now see what happens when each of the reactants is bifunctional. Therefore, (a) equals A–R–A and (b) equals B–R′–B. Now, each of these species has two groups that are capable of reacting with one another. For our schematic representation of the reaction we can write

$$\underset{(a)}{A{-}R{-}A} + \underset{(b)}{B{-}R'{-}B} \longrightarrow \underset{(c)}{A{-}R{-}A{-}B{-}R'{-}B} \tag{3}$$

We now find, contrary to our previous experience, that the product is also bifunctional. Hence further reaction, with the concomitant building up of the molecule or chain, can take place. For example, we can have

$$\underset{(a)}{A{-}R{-}A} + \underset{(c)}{A{-}R{-}A{-}B{-}R'{-}B} \longrightarrow A{-}R{-}A{-}B{-}R'{-}B{-}A{-}R{-}A \tag{4}$$

The new species formed is still bifunctional and could react further with (b). Another possibility would be for (c) to react with itself so that

$$\underset{(c)}{A{-}R{-}A{-}B{-}R'{-}B} + \underset{(c)}{A{-}R{-}A{-}B{-}R'{-}B} \longrightarrow A{-}R{-}A{-}B{-}R'{-}B{-}A{-}R{-}A{-}B{-}R'{-}B \tag{5}$$

Here again the bifunctionality is maintained in the product so that further reaction can take place. Therefore, with each successive step more reaction pathways develop. These will soon become much too numerous to attempt to describe in detail. However, it should be abundantly clear that in this stepwise manner very long chain molecules can be formed. The length of each chain will not be the same, and consequently there will be a distribution of molecular sizes in the resulting product. This distribution of sizes is a unique feature of polymer synthesis. In more classical studies uniformity in molecular weight of the species formed is the rule. For the more quantitative analysis of polymerization, however, we must be concerned with molecular weight averages.

Many important classes of polymers are in fact synthesized by the method just outlined. This mode of synthesis is called condensation polymerization. For example, in analogy with Eq. (3), if our initial reactants are a dialcohol such as ethylene glycol, $HOCH_2CH_2OH$, and a diacid such as adipic acid, $HO\text{--}\overset{O}{\overset{\|}{C}}\text{--}\underset{H}{\overset{H}{C}}\text{--}\underset{H}{\overset{H}{C}}\text{--}\underset{H}{\overset{H}{C}}\text{--}\underset{H}{\overset{H}{C}}\text{--}\overset{O}{\overset{\|}{C}}\text{--}OH$ in the first step of the polymerization,

$$HO\text{--}\overset{O}{\overset{\|}{C}}\text{--}CH\text{--}CH\text{--}CH\text{--}CH\text{--}\overset{O}{\overset{\|}{C}}\text{--}OH + HOCH\text{--}CH\text{--}OH \longrightarrow$$

$$HO\text{--}CH\text{--}CH\text{--}O\text{--}\overset{O}{\overset{\|}{C}}\text{--}CH\text{--}CH\text{--}CH\text{--}CH\text{--}\overset{O}{\overset{\|}{C}}\text{--}OH + H_2O \tag{6}$$

water is given off and an ester is formed. This ester contains an OH (hydroxyl) and a $\overset{O}{\overset{\|}{C}}\text{--}OH$ (carboxyl) end group. Further stepwise reactions can take place, and a polymer will be formed. In this example a linear polyester of the type given in Table 2.1 is synthesized. When a diamine such as hexamethylene diamine $H\text{--}\overset{H}{N}\text{--}\underset{H}{\overset{H}{C}}\text{--}\underset{H}{\overset{H}{C}}\text{--}\underset{H}{\overset{H}{C}}\text{--}\underset{H}{\overset{H}{C}}\text{--}\underset{H}{\overset{H}{C}}\text{--}\underset{H}{\overset{H}{C}}\text{--}\overset{H}{N}\text{--}H$ is substituted for the dialcohol, a polymide will be formed following the same synthetic scheme. Here water is formed by the reaction of H from the amine with the OH group of the acid. When a dialcohol and a diisocynate, $O{=}C{=}N\text{--}R\text{--}N{=}C{=}O$, react, then a polyurethane is synthesized.

A very important concept is embodied in the above discussion. We have found, in effect, that the principles of organic chemistry that are involved in synthesizing the polymers are exactly the same as those applicable to monomeric systems. The fact that we are dealing with very high molecular weight species has in no way altered the basic synthetic organic chemistry that is involved. Many different kinds of condensation reactions are known. In addition to the typical synthetic polymerizations cited above, these also include natural occurring reactions, which take place in cells and yield such diverse substances as cellulose, starch, other polysaccharides, the proteins, and polynucleotides. The natural occurring polymerizations, being of a highly specific character, usually yield species of uniform composition and size.

Another major method of synthesis is termed addition polymerization. It is the mode by which ethylene $H_2C{=}CH_2$, propylene $H_2C{=}CHCH_3$, and vinyl monomers such as styrene $H_2C{=}CH\text{--}C_6H_5$ are polymerized to their respective homopolymers. When we look at the structure of the ethylene monomer, $\overset{H}{\underset{H}{}}\!\!>C{=}C<\!\!\overset{H}{\underset{H}{}}$

which can serve as a prototype for others, it would appear at first glance, that the necessary requirement of bifunctionality is not met. However, a more detailed examination and insight into the nature of the carbon-carbon double bond shows that this discrepancy is more apparent than real. The two electron pairs that form the double bond are not of the same strength.[3] One pair of electrons, or one bond, will be much more reactive than the other. Hence, although it is not obvious in its chemical formula, our monomer inherently possess bifunctional character. To take advantage of the carbon-carbon double bond of monomers a catalyst is required for an initiation step. A very convenient catalyst is a free radical. This is a species with an unshared electron which can be represented as R•. Schematically, the polymerization proceeds as follows: In the first step, initiation, a monomer unit has become

$$\mathrm{R\cdot} \;+\; \begin{array}{ccc} \mathrm{H} & & \mathrm{H} \\ \diagdown & & \diagup \\ & \mathrm{C::C} & \\ \diagup & & \diagdown \\ \mathrm{H} & & \mathrm{X} \end{array} \longrightarrow \begin{array}{ccc} & \mathrm{H} & \mathrm{H} \\ & | & | \\ \mathrm{R:} & \mathrm{C:} & \mathrm{C\cdot} \\ & | & | \\ & \mathrm{H} & \mathrm{X} \end{array} \qquad (7)$$

a free radical. It, in turn, can now react with other monomer units to form a polymer by the following scheme:

$$\begin{array}{ccc} & \mathrm{H} & \mathrm{H} \\ & | & | \\ \mathrm{R-} & \mathrm{C-} & \mathrm{C\cdot} \\ & | & | \\ & \mathrm{H} & \mathrm{X} \end{array} \;+\; \begin{array}{ccc} \mathrm{H} & & \mathrm{H} \\ \diagdown & & \diagup \\ & \mathrm{C::C} & \\ \diagup & & \diagdown \\ \mathrm{H} & & \mathrm{X} \end{array} \longrightarrow \begin{array}{ccccc} & \mathrm{H} & \mathrm{H} & \mathrm{H} & \mathrm{H} \\ & | & | & | & | \\ \mathrm{R-} & \mathrm{C-} & \mathrm{C-} & \mathrm{C-} & \mathrm{C\cdot} \\ & | & | & | & | \\ & \mathrm{H} & \mathrm{X} & \mathrm{H} & \mathrm{X} \end{array} \qquad (8)$$

The initiating or activating species is said to have opened up the double bond. As this propagation step, Eq. (8), is continued, hundreds to thousands of monomer units can be added to the growing chain. The center of activation or reactive site is at the end of the growing chain. Termination of the growing chain will occur when two free radicals combine. There are also other possible modes of termination. Thus, addition polymerization is essentially a chain reaction. A small number of monomers become activated. Once started, chain growth, or monomer addition, can proceed very rapidly. One initial event can produce a polymer chain containing the order of 10,000 monomer units. In the case illustrated it proceeds by a free radical mechanism. By using different kinds of initiators the polymerization can also proceed by an ionic mechanism.

There are some very important differences between the two major polymerization processes that we have been discussing. In condensation polymerization the reaction between two growing polymer chains leads to a further enhancement of the molecular weight. But in addition polymerization, reaction between two growing chains leads to a cessation of further growth as far as the two molecules involved are concerned. In addition polymerization an individual polymer molecule, once initiated, is synthesized very rapidly from the unreacted monomer pool. However, to obtain a good total yield of polymer requires several hours. Very high molecular

[3] A student of orbital theory will recognize that we are discussing σ and π bonds.

weights can be obtained quite routinely from this type of chain reaction. In condensation polymerization, the molecular weight is built up by a series of independent reactions proportioned over the total time period of the polymerization. At the early stages of the process, the average molecular weight of the system will be low, although the amount of unreactive material present will be negligible. As the reaction continues, the average molecular weight will also increase. However, to obtain very high molecular weights, the reaction will have to proceed virtually to completion. Molecular weights greater than 25,000 are rare for a linear condensation polymer. In contrast, with addition polymerization very high molecular weights are found at the very early stages of the process. Continuation of addition polymerization is dictated by the amount of polymer required rather than an effort to obtain higher molecular weights.

From these simple considerations of the two basic polymerization mechanisms we have found that the polymer molecules formed will not all have the same molecular weight, or chain length. Usually, broad molecular weight distributions result. The constitution of the system must then be described either by a set of different average molecular weights or by the distribution function itself. Many properties of polymeric systems depend on the details of the molecular weight distribution. Properties of chemically identical systems can be quite different depending upon whether this distribution is narrow or broad. Examples of typical distribution functions, for a condensation polymer, are given in Figure 2.15, for the case where 95%

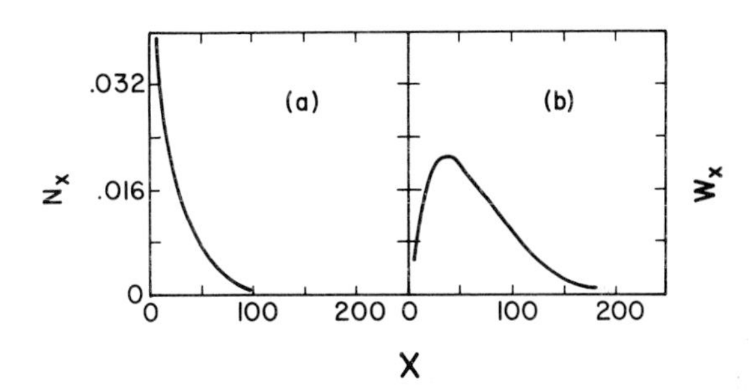

Figure 2.15. Molecular weight distribution functions for condensation polymers. (a) Number fraction as a function of chain length. (b) Weight fraction as a function of chain length. Adapted from Paul J. Flory, *Principles of Polymer Chemistry*. Copyright 1953 by Cornell University Press. Reproduced with permission.

of all the molecules have reacted. Figure 2.15(a) gives the relative number of molecules of a given length that have formed plotted against the number of repeating units, x. On a number basis the monomer is the most predominant species, followed by the dimer, trimer, etc. When considered on a weight basis, however, as is illustrated in Figure 2.15(b), the distribution has a maximum and is characterized by much higher molecular weights. Many of the addition-type polymerizations yield very similar molecular weight distributions.

Branching, Cross-Linking, and Network Formation

The polymerization schemes that have been outlined lead to the formation of linear chains in the absence of any complicating side reactions. To form branched structures, at least one of the monomeric units must have a functionality greater than two. If, for example, a trifunctional acid reacted with a difunctional alcohol, a highly ramified structure, of the kind illustrated in Figure 2.5 would result. Many nonlinear polymers of interest are composed mainly of bifunctional units. The units with a higher functionality are usually a minor constituent and randomly distributed. In addition polymerization, branching and network formation can also be accomplished by including a small amount of a divinyl monomer, such as divinyl benzene $H_2C{=}CH{-}C_6H_4{-}CH{=}CH_2$, in the polymerization mixture.

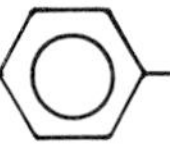

Celluloid, which is nitrated cellulose, is generally credited with being the first plastic. A naturally occurring macromolecule is the starting material in this case. However, since celluloid is highly flammable, a great deal of effort was made to find a substitute material. The first completely synthetic plastic was prepared by Baekeland in 1907 and is called Bakelite. It still has a widespread use despite the hundreds of polymers that have been synthesized since. Bakelite is not, however, a linear chain. It is prepared from the reaction of phenol $C_6H_5{-}OH$ with formaldehyde $H{-}C(H){=}O$ to yield a three-dimensional network polymer as is indicated in Figure 2.16. This polymer is a very strong and tough, but lightweight, material that is easily molded into a variety of objects.

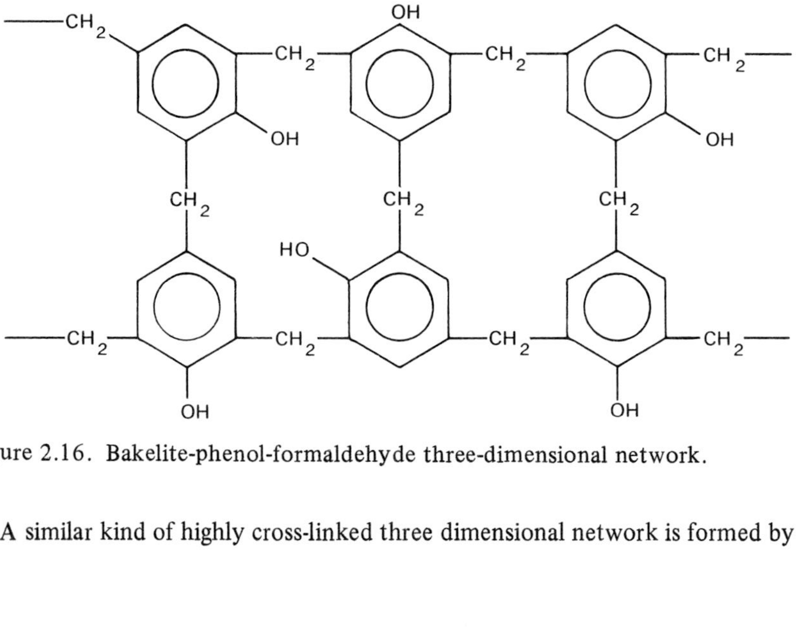

Figure 2.16. Bakelite-phenol-formaldehyde three-dimensional network.

A similar kind of highly cross-linked three dimensional network is formed by the reaction of formaldehyde and urea, $O{=}C(NH_2)_2$, yielding a polymer with the structure as indicated in Figure 2.17. The reaction of formaldehyde with the trifunc-

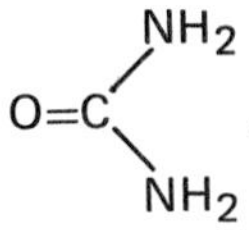

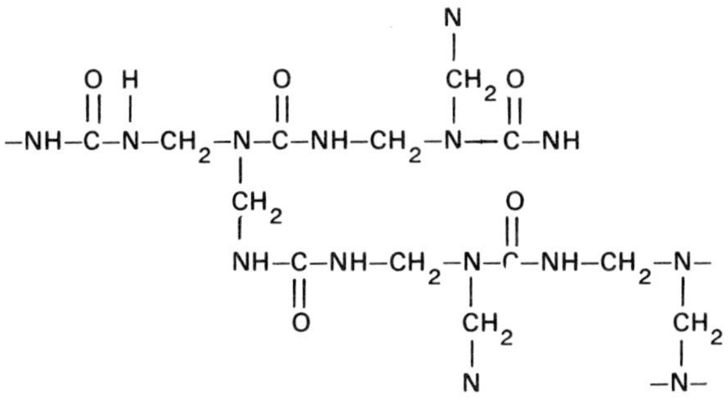

$$
\begin{array}{c}
\text{N} \\
| \\
\text{CH}_2 \\
| \\
-\text{NH}-\overset{\overset{\text{O}}{\|}}{\text{C}}-\overset{\overset{\text{H}}{|}}{\text{N}}-\text{CH}_2-\text{N}-\overset{\overset{\text{O}}{\|}}{\text{C}}-\text{NH}-\text{CH}_2-\text{N}-\overset{\overset{\text{O}}{\|}}{\text{C}}-\text{NH} \\
\text{N}(-\text{CH}_2-\text{NH}-\underset{\underset{\text{O}}{\|}}{\text{C}}-\text{NH}-\text{CH}_2-\underset{\underset{\text{N}}{|}}{\underset{\text{CH}_2}{\text{N}}}-\overset{\overset{\text{O}}{\|}}{\text{C}}-\text{NH}-\text{CH}_2-\underset{\underset{-\text{N}-}{|}}{\underset{\text{CH}_2}{\text{N}}}-)
\end{array}
$$

Fig. 2.17. Urea-formaldehyde polymer.

tional amine melamine

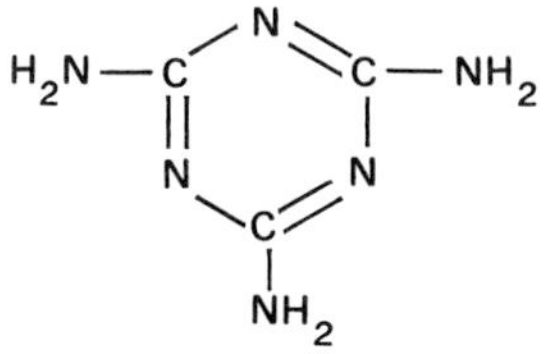

$$C_3N_3(NH_2)_3 \quad \text{(1,3,5-triazine ring with } -NH_2 \text{ on each C)}$$

also yields a highly cross-linked network polymer. The melamine polymers are used extensively in lightweight dinnerware and as laminates for counter and table tops.

The epoxies are another example of highly cross-linked systems. They are widely known as exceptional adhesives, and also find uses as surface coatings and in the encapsulation of electronic components. The outstanding features of these polymers are toughness and excellent adhesive or bonding properties with many type surfaces. Although the chemical structures of the different epoxies available vary in detail, they generally result from the reaction product of epichlorohydrin

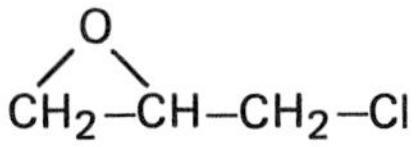

$$\overset{\;\;O\;\;}{\widehat{CH_2-CH}}-CH_2-Cl$$

and bisphenol A

$$HO-C_6H_4-\underset{\underset{CH_3}{|}}{\overset{\overset{CH_3}{|}}{C}}-C_6H_4-OH$$

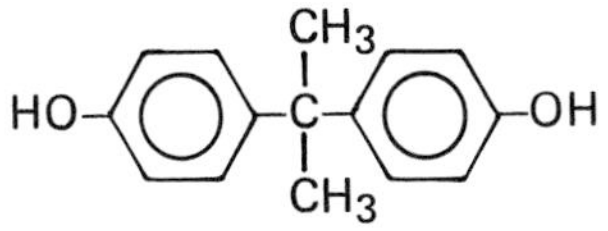

to yield the chain

$$\overset{\;\;O\;\;}{\widehat{CH_2-CH}}-CH_2-\left[O-C_6H_4-\underset{\underset{CH_3}{|}}{\overset{\overset{CH_3}{|}}{C}}-C_6H_4-O-CH_2-\overset{\overset{OH}{|}}{CH}-CH_2-O-\right]_n$$

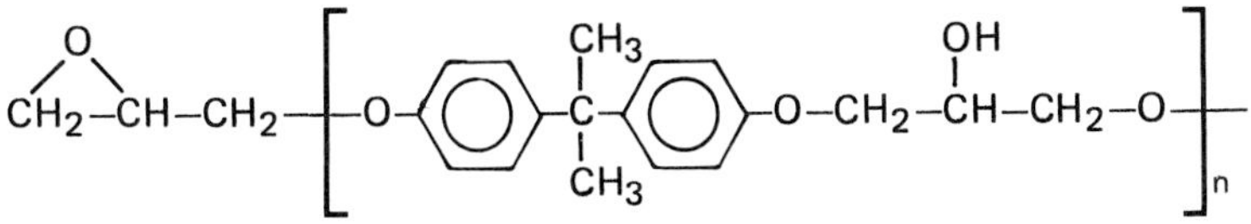

These chains are subsequently intermolecularly cross-linked to form the final product. Depending on the value of n, the product can vary from being a viscous liquid to a rigid solid. This property is controlled by the initial ratio of epichlorohydrin to *bis*-phenol. Cross-linking reactions are carried out, either through the epoxy group, ($\overset{O}{CH_2{-}CH}{-}$), or the hydroxy group, (OH). Amines are very effective in rapidly forming intermolecular cross-links between the epoxy groups: Very tough, hardened systems result which strongly resist heat distortion.

As has been illustrated with the epoxies, three-dimensional networks can be formed from a collection of linear chains by the introduction of intermolecular cross-links after the polymers have been formed. This procedure requires having reactive groups in the chain and a cross-linking reagent that has at least bifunctional character, so that it can react with at least two such units. More than a century ago, Charles Goodyear discovered the vulcanization of rubber, which is the classical intermolecular cross-linking process. In Chapter 4 the importance of this chemical process to the elastic properties of natural rubber will be discussed in detail. Much less than one percent of the isoprene units are intermolecularly cross-linked in a typical rubber elastic system. When the cross-linking level is increased to the order of several percent, what is commonly called a hard rubber results.

The tanning of leather is another manifestation of the intermolecular covalent cross-linking on properties. Tanning is merely the intermolecular cross-linking of the fibrous protein collagen, the major constituent of leather. Here the cross-linking process increases the thermal stability, reduces the shrinkage, and renders the material less susceptible to the action of solvents, particularly water. Wool and hair owe many of their physical properties to the presence of sulfur-sulfur intermolecular cross-links between polypeptide chains.

Another method of cross-linking, which is applicable to many polymeric systems, involves subjecting the chains to the action of high energy ionizing radiation from neutrons or gamma rays. In the case of polyethylene, for example, this high energy source strips an occasional hydrogen atom from the chain and leaves a free radical behind. The free radicals from different chains can combine with each other to form intermolecular covalent cross-links. This method of cross-linking has been used extensively with polyethylene. Since the cross-linking process retards flow, objects can be produced that retain their shapes at much higher temperatures than would be usual. This is an extremely useful property for branched polyethylene when sterilization in boiling water is necessary. Caution needs to be exercised when this process is used, since for some polymers the radiation can also cause chain scission. In these cases the molecular weight will be reduced with concomitant deliterious effects.

In this brief introductory survey we have seen the diversity of structures and chemical types that can be manifested in macromolecules. Concomitantly, there is a wide range of properties that can be attained. These, in turn, direct and govern the end-use of synthetic polymers and the biological function of the naturally occurring ones. Although we have accepted the fact that such a wide range in properties exists, we have not as yet inquired as to what factors cause the differences. Put another way, the question can be asked as to why one type of chain makes a good elastomer while another is a tough inextensible plastic. Why can genetic information be stored in nucleic acids? Or why does nylon make such a good fiber comparable to nature's own silk and wool? What is it that makes the recently commercialized aramid fibers so strong? We seek the answers to these and similar questions in the following chapters by first examining the structure of polymer chains in more detail and then relating the structure to properties.

Chapter 3

CHAIN STRUCTURE

Internal Rotation

The pioneering work over the past 30 to 40 years by such scientific luminaries as Staudinger, Svedberg, Debye, and Flory—all Nobel Laureates—has established beyond doubt that molecules of very high molecular weight exist. By use of modern experimental and theoretical methods, a quantitative description that is the basis for the understanding of the properties of such molecules has been developed. The usual or normal laws of chemistry and physics were found to apply equally well to this class of molecules. However, in many cases, a generalization of the existing laws was necessary in order to include monomeric and polymeric systems in a unified way. This unification has led in fact to a better understanding of the underlying basis for these laws. Thus, besides resulting in a very useful class of molecules, the study of polymers has given us a better and much broader insight into some very important scientific principles.

Establishing the fact that we are dealing with molecules of very high molecular weight is obviously the first step to an understanding of their properties. However, this knowledge is not in itself sufficient to explain the diversity of properties characteristic of macromolecular systems. Another major factor is the structure or spatial form that these molecules can assume. It should be intuitively obvious that a chain molecule of high molecular weight will possess a great deal of structural versatility. In the ultimate analysis it is this structural versatility, and the consequences thereof, that distinguishes macromolecular systems from monomeric ones. We shall examine these structural features in great detail as we seek the underlying basis for the properties of macromolecules.

A linear chain has already been given a schematic representation in Figure 2.1. The symbol A can be replaced by an explicit chemical formula such as—CH_2—so that a polyethylene chain is being described. A more detailed representation of this chemically simple chain can be given as shown in Figure 3.1. In this representation we have precisely fixed the length of the C—C bond at its accepted value of 1.53 Å and have also fixed the valence angle at its value of 109°. Even when the normal bond lengths and angles are specified, the representation given in Figure 3.1 is not adequate. The reason for this is that we have of necessity restricted ourselves to a structure, or conformation as it is called, that is constrained to lie in the plane of the paper. This is an arbitrary restriction on our part and occurs because the written word is conventionally displayed in two-dimensional space. In writing this structure we have not allowed for the possibility of rotations about the single bonds that make up the chain. Such rotations can be accomplished while the normal valence angles and bond lengths are maintained. This capability for internal rotation allows a chain molecule to generate a large number of different conformations. Conformations are defined as the nonidentical arrangements produced by the rotation of atoms around

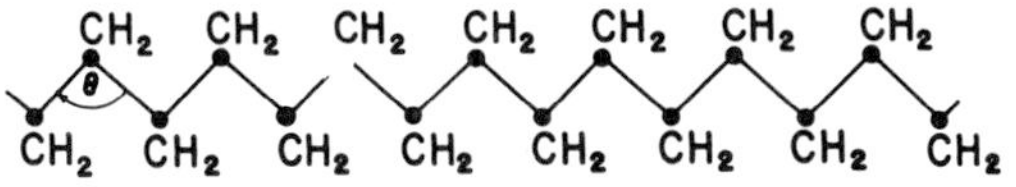

CH₂ CH₂ CH₂ CH₂ CH₂ CH₂ CH₂

θ

CH₂ CH₂ CH₂ CH₂ CH₂ CH₂ CH₂ CH₂

Figure 3.1. Polyethylene chain.

one or more single bonds. The conformation illustrated in Figure 3.1 is thus one that has conveniently, but arbitrarily, been selected from among many.

We must now ask what we mean by internal rotation and how this operation can lead to many different structures. The fundamental principle involved here can be easily understood by studying a simple molecule such as ethane C_2H_6. This molecule contains but one rotatable bond, the carbon-carbon single bond. A representation of two conformations of ethane is given in Figure 3.2. At first sight these two conformations might appear to be the same. However, close examination shows that they are indeed different. The three hydrogen atoms attached to each of the carbons are arranged differently relative to one another. In Figure 3.2(a) each set of hydrogen atoms is directly opposite each other, as indicated by the vertical dashed lines. This is known as the eclipsed conformation. If one carbon atom is now rotated relative to the other, along the axis of the C–C bond, the relationship between the two sets of hydrogen atoms will be changed. When a rotation of 60° is made about the C–C bond the structure represented in Figure 3.2(b) results. Here, the hydrogen atoms from each set are as far from one another as is possible. This is known as the staggered conformation. The staggered conformation will be repeated for every

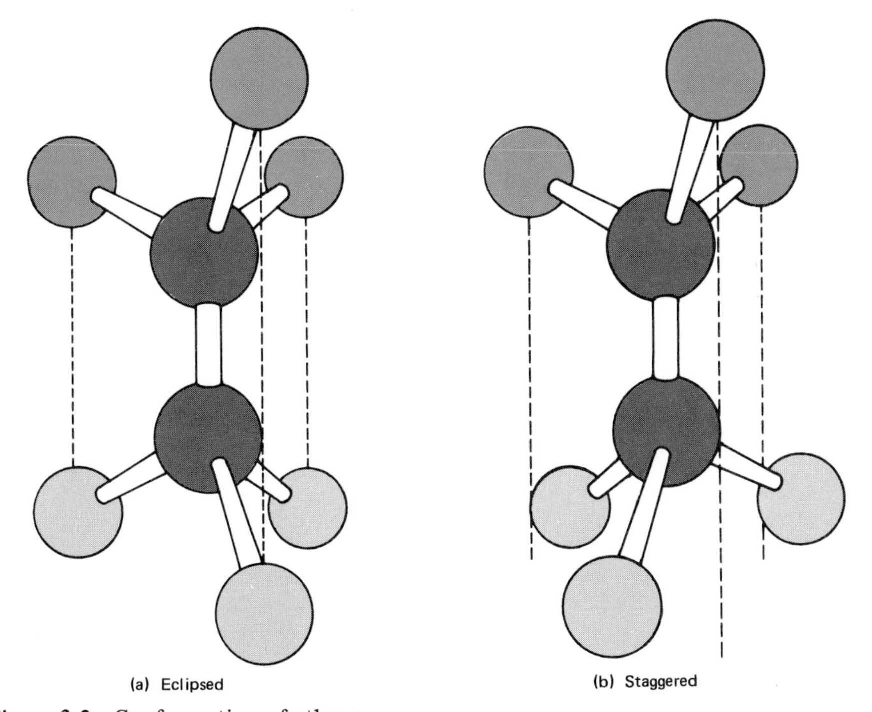

Figure 3.2. Conformation of ethane.

120° of rotation, so that there are three such structures. However, since each of the hydrogen atoms is in exactly the same relative position, for each rotation of 120° from that of Figure 3.2(b), identical, or equivalent, structures result.

From a purely operational point of view, an infinite number of conformations can be generated by rotation about the carbon-carbon bond. However, one should not conclude from this that all these possible conformations are equally probable. The probability that a conformation will exist depends on its potential energy. Conformations characterized by low energy will be found much more frequently than those with very high energy. In the eclipsed conformation of Figure 3.2(a), the hydrogen atoms from each set are as close to each other as the geometric conditions allow. They are, in fact, placed so close to one another that strong repulsive forces arise. These repulsive forces make the energy of the eclipsed form much greater than that of the staggered one. Hence, it will be a very much less favored structure. Following arguments of this kind, we can represent the potential energy of ethane as a function of the rotational angle ϕ, as in Figure 3.3. Each of the equivalent staggered forms are arbitrarily assigned rotational angles 0°, +120°, and -120°, respectively. These are the conformations of lowest energy, since the hydrogen atoms are as far apart from one another as possible. The highest energy, eclipsed forms, are displaced from each of these by rotations of 60°. The energies characteristic of intermediate angles of rotation lie between the two extremes. We see, therefore, that although all rotational angles are geometrically possible in ethane, certain ones are pre-

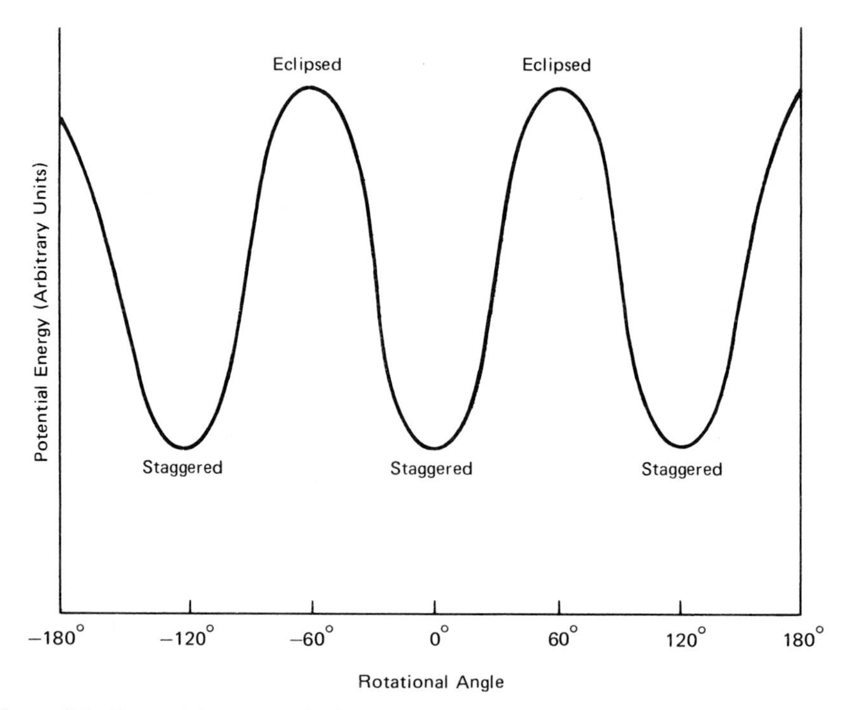

Figure 3.3. Potential energy of ethane as a function of rotational angle.

ferred on energetic grounds because of the interaction between the hydrogen atoms. These atoms are not involved in chemical bonds and are called nonbonded atoms.

The rotational freedom about a single bond is a result of many factors. A very important contribution to this rotation comes from the nature of the bond itself. Most commonly found in chain molecules are the carbon-carbon, carbon-oxygen, silicon-oxygen, carbon-nitrogen, single bonds, and more recently phosphorous-nitrogen bonds. Such bonds have an inherent rotational or torsional freedom that depends on the details of its electronic structure. Double bonds such as are found in ethylene and in dienes have very little rotational freedom and can be thought of as being rigid or nonrotatable. The inherent rotational characteristics of a bond will be modified by the groups that are attached to the chain atoms, called the substituents. The interactions between the nonbonded atoms will restrict the preferred rotational angles. The size of these atoms, or their steric influence, is very important in this respect—as we have already seen even for the small hydrogen atoms. When bulkier substituents are attached, the rotational freedom is further restricted. The steric influences are modified further by the repulsive and attractive forces between the atoms.

A somewhat more complicated situation arises when the internal rotation in *n*-butane is analyzed. This molecule can be represented in skeletal form as shown in Figure 3.4. We focus attention here on the rotation about the central bond that con-

$$H_3C-CH_2-CH_2-CH_3$$

1 2 3 4

Figure 3.4. Skeletal representation of *n*-butane.

nects carbon atoms 2 and 3. The substituents on these two carbon atoms are now no longer identical. In relation to ethane, a hydrogen atom from each of the carbon atoms has been replaced by the much bulkier $-CH_3$ group. The staggered conformations will clearly still be more favored than the eclipsed ones. However, in this case, not all of the three staggered conformations will be energetically equivalent. In one of the staggered conformations, the two terminal methyl groups will be as far apart from one another as possible. This structure is shown in Figure 3.5 and is the planar

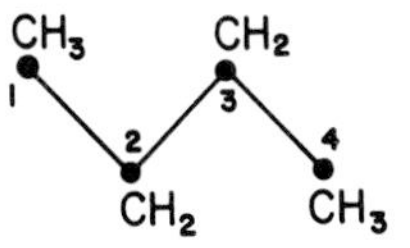

CH_3 CH_2

1 2 3 4

CH_2 CH_3

Figure 3.5. *Trans* conformation *n*-butane.

or *trans* form. It is the conformation of lowest energy. Either a clockwise or a counterclockwise rotation of 120° about the central bond will also yield a staggered conformation as is shown in Figure 3.6. These two structures, which are called the *gauche* form are skewed or nonplanar. The distance between the two CH_3 groups is the same in both of these forms. These two groups are, however, closer to one another than in the planar *trans* form. In the *gauche* forms, the hydrogen atoms attached to the terminal carbons will repel one another because of the closeness of approach. Consequently, the energy of the *gauche* forms will be somewhat greater

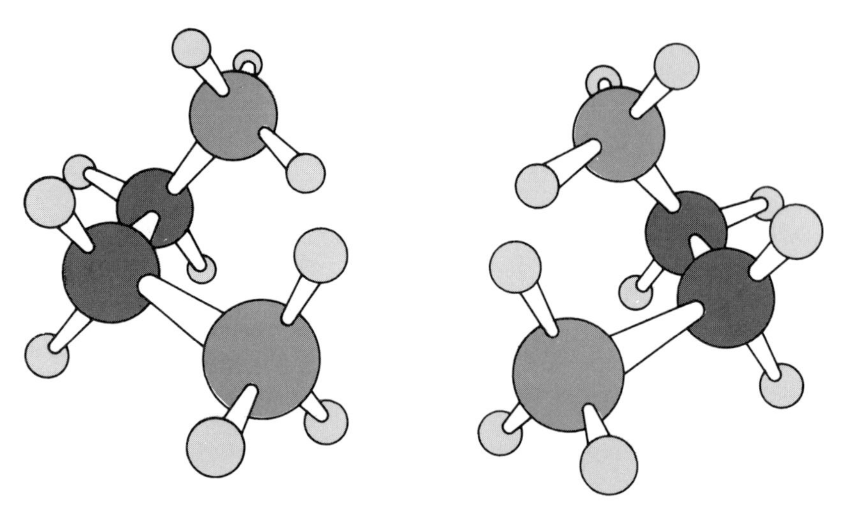

Figure 3.6. The two *gauche* conformations of *n*-butane.

than that associated with the *trans* form. The occurrence of *gauche* conformations will, therefore, be slightly less favored. At ordinary temperatures about half as many molecules will be found in the *gauche* form as compared to the *trans*.

The relative energy of the *n*-butane conformations as a function of rotation about the internal carbon-carbon bond is given in Figure 3.7. Here, the most energetically favored *trans* conformation is assigned the rotational angle $\phi = 0$. The two *gauche* conformations are then located at $\pm 120°$, respectively. Although their energies are higher than the *trans* form, the fact that they are still a favored conformation is represented by the minimum in the energy function. If we compare this diagram with the corresponding one for ethane (Figure 3.3), we see that the favored conformations still occur at the same rotational angles for both molecules. However, for ethane the three favored or staggered conformations each possess the same energy.

The concept of internal rotation, as was illustrated by the discussion of *n*-ethane and *n*-butane, can now be applied to the long-chain polyethylene molecule. Let us consider a set of three successive bonds centrally located within the chain as shown in Figure 3.8. The first two bonds, connecting atoms C_1 to C_2 and C_2 to C_3, form a plane that we shall take as the plane of the paper. The normal valence angle is designated by θ. Atom C_4 must be located on the indicated circle, which is the base of the cone described by the rotation of bond 3 while the valence angle is kept at θ. The exact position of atom C_4 on this circle will depend on the angle of rotation of bond 3 about bond 2. If all rotational angles were equally probable, then each position on the circle would also be equally probable. In this case, the bond is said to be freely rotating. The analysis of *n*-butane has indicated that this is not a realistic concept and will therefore not apply for a bond in the polyethylene chain. Here, three preferential but not equally probable rotational angles or rotational states can be chosen. In Figure 3.8, for example, position C_4' represents the *trans* conformation while C_4'' is one of the *gauche* forms. One can continue in this manner and locate the successive bond termini along the chain. The conformation, and thus the spatial form of the long-chain molecule is determined by specifying the rotational angle of each

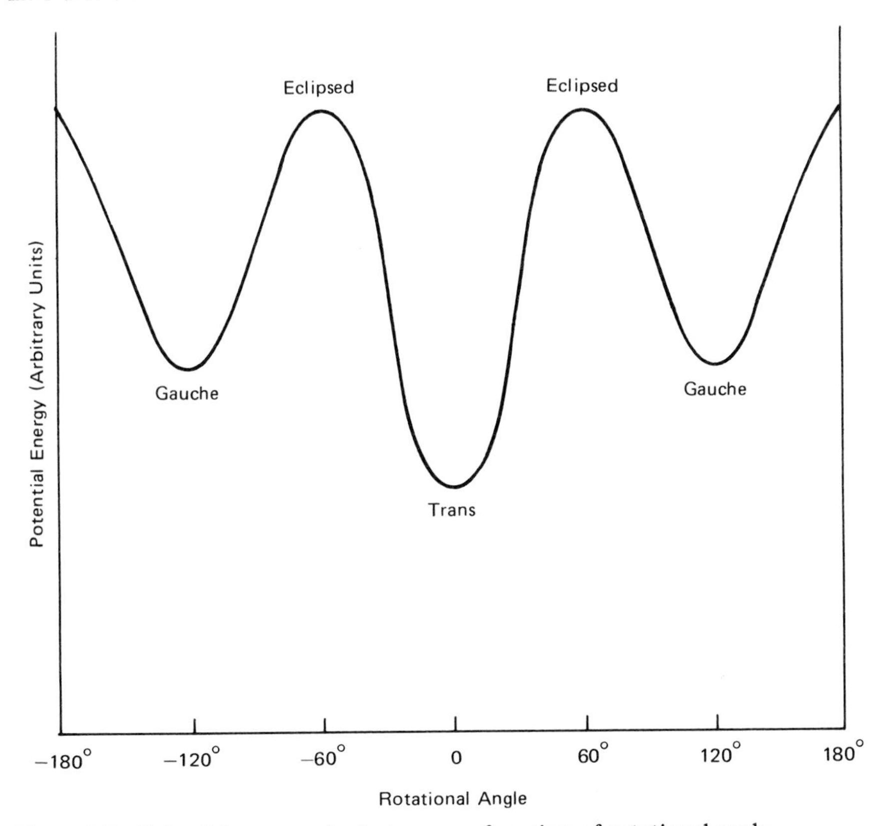

Figure 3.7. Potential energy of *n*-butane as a function of rotational angle.

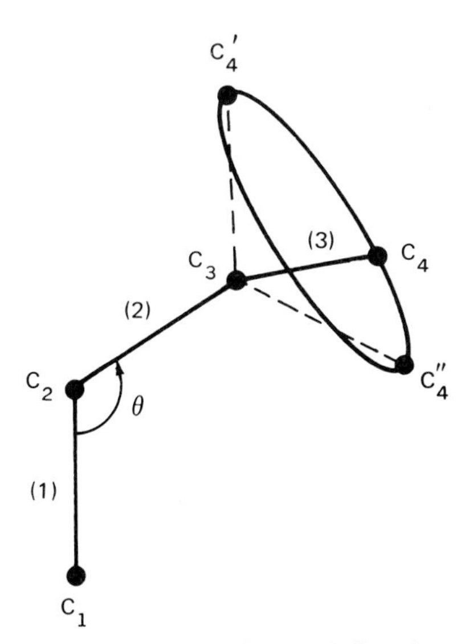

Figure 3.8. Spatial representation of single bonded carbon chain. Reprinted from Paul J. Flory, *Principles of Polymer Chemistry*, Copyright 1953 by Cornell University Press. Used by permission of Cornell University Press.

of the constituent single bonds. The molecule that is formed must clearly have a highly irregular form and shape and cannot be properly represented in any planar form.

Disordered Chain Conformation

For a chain of n bonds, n-2 rotational angles about the single bonds need to be specified in order to describe the conformation. However, altering the rotational position of only one bond, among the many thousand, will change the conformation. Such alterations easily occur as a consequence of normal thermal motions. Therefore, we are faced with the problem not only of specifying a given conformation but also of enumerating all the conformations that are available to a long-chain molecule. To estimate the number of conformations that a chain molecule can assume, we will allow each bond to adopt independently one of the three preferred rotational angles. If there are n bonds per chain, then the molecule can assume 3^n distinct conformations. If n = 10,000, which represents a modest size chain, then 3^n is equal to $10^{4,771}$. This is a number of such astronomical proportion as to defy comprehension. As a frame of reference, it can be compared with Avogadro's number, the number of molecules in a mole. This number is a mere 10^{23} as compared to the above. It is extremely unlikely, therefore, that two molecules would be found in the same conformation. We thus have our first insight into the tremendous structural and conformational versatility possessed by long-chain molecules. Not only is the molecular shape highly irregular but there are also an extraordinarily large number of possible conformations. The principles we have used in analyzing the structure of polyethylene will hold equally well for other kinds of long-chain molecules. Usually, only minor modifications concerned with the number, position, and energies of the preferred rotational angles are necessary.

A schematic and, by necessity, two-dimensional representation of one particular structure that can be generated by the procedure outlined above is shown in Figure 3.9. The arrow represents the vector distance $\vec{r}$ from one end of the chain to the other. Another pictorial method of representing a chain conformation is using molecular models as is illustrated in Figure 3.10. This model represents a polyethylene conformation for a chain consisting of 150 carbon atoms. In both these models, the tortuous, random-like structure and highly irregular shape of the molecule becomes readily apparent. The molecule is highly coiled and very compact, since the end-to-end distance $\vec{r}$ is very much less than the extended, or contour, length of the chain. In contrast, the planar zigzag structure illustrated in Figure 3.1 yields an extended molecule of rod-like form. However, it represents but one of the large number of conformations that can be generated. The overwhelming majority of these will be characterized by their compactness, irregularity, and lack of any definite order. In popular usage molecules having such structures have been called random coils. More properly, perhaps, they should be termed statistical conformations.

At first glance the overwhelming number of conformations available to each chain would appear to make for a rather hopeless and even ridiculous situation in which to attempt to relate the properties of macromolecules to their structure. This task would be particularly difficult if we were to use the methods conventionally

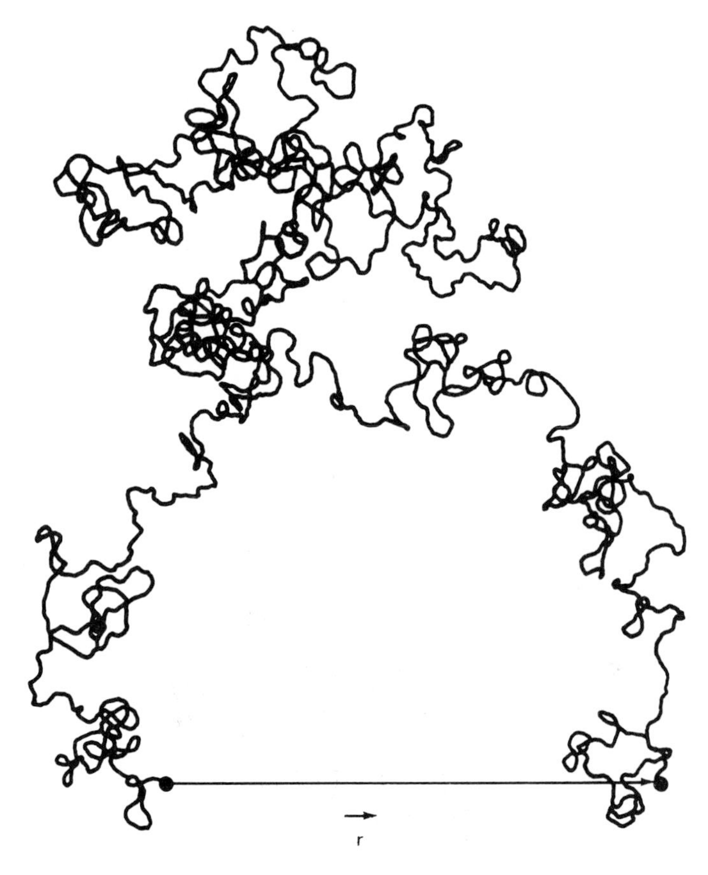

Figure 3.9. One conformation of a 1000 link polyethylene chain. Adapted from L. R. G. Treloar, *The Physics of Rubber Elasticity*, The Clarendon Press, Oxford. Reproduced by permission.

employed in studying the properties of monomeric substances. For, if these procedures were adopted, we would be required to locate each atom precisely in space. To do this for all possible chain conformations would exceed the capacity of our highest speed computer. Furthermore, even if the necessary data could be assembled, they would be so numerous as to be absolutely useless for any practical purposes. Fortunately, we are concerned only with properties that can be measured. Physical measurements represent the average value of a given property for a collection of many molecules and thus are averaged over many conformations. Hence, we need to be concerned only with the average properties of our chain molecules rather than with a detailed geometric description of each structure. It is exactly the difficulty that we have mentioned, namely the large number of conformations involved, that makes polymer systems ideally suited to apply statistical methods in order to obtain average values.

Another question is what characteristic of a chain deserves our attention for the purpose of calculating its average value. Although there are a large number of

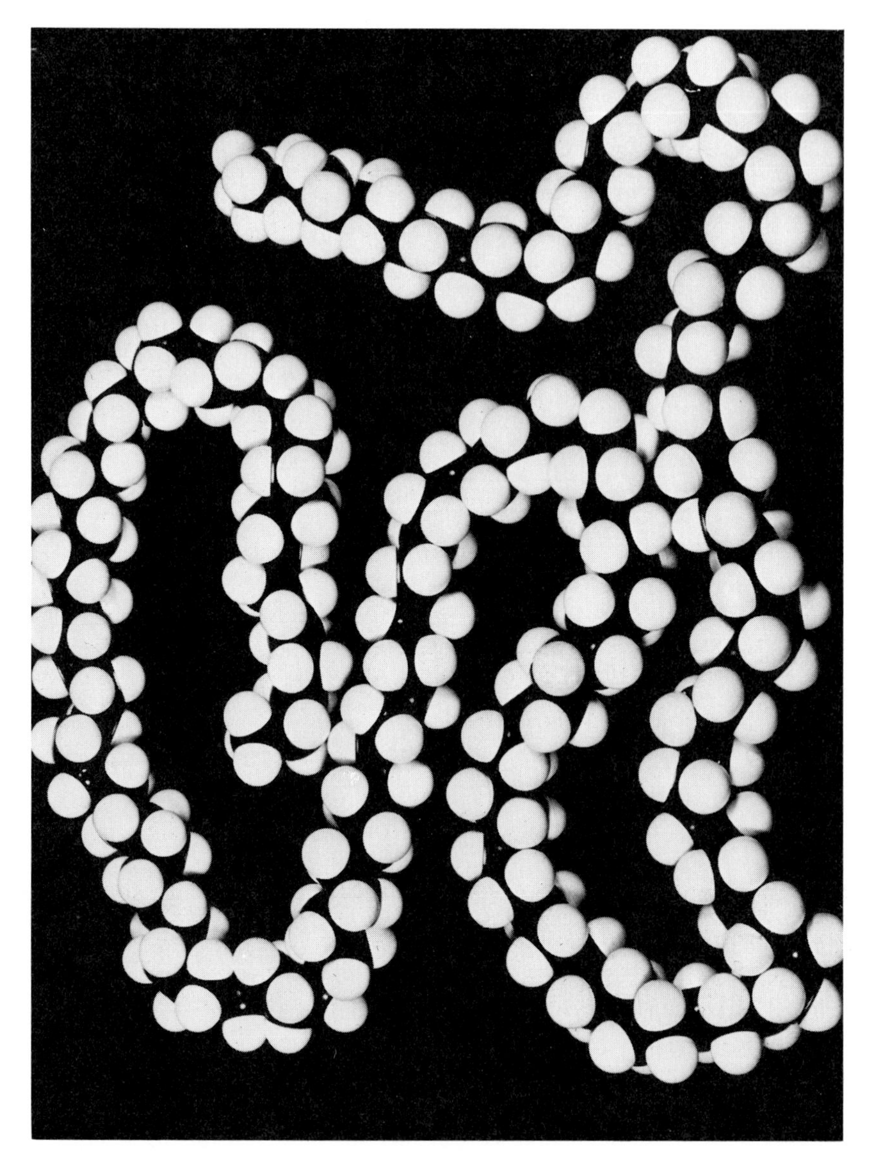

Figure 3.10. Molecular model representation of a conformation of a 150 carbon polyethylene chain.

choices, it turns out that a quantitative description of the size, or dimension, of the chain is of prime importance. A very convenient measure of the size of a chain, or its spatial characteristics, is the average of the square of r, which we shall represent as $\langle r^2 \rangle$.[1] Many properties of the chain depend on this average value. In principle this average quantity can be obtained by calculating r for each conformation, mul-

[1] We should recall that r represents the end-to-end distance of a chain (see Figure 3.9).

tiplying its square by the probability of the occurrence of the particular conformations and taking the sum of this product overall the possible conformations. A direct enumeration of the desired average quantity is clearly an impossible task because of the multitudinous number of conformations involved. Fortunately, mathematical methods have been developed which avoid the necessity for direct enumeration, but still allow for an exact calculation of $\langle r^2 \rangle$ for all real chains. The only molecular information that needs to be supplied for these calculations is the number and type of bonds and the relative energies that are associated with the preferred rotational angles. The details of how these calculations are carried out are well beyond our present scope. However, we can state that the results of the exact calculations are in extraordinarily good agreement with experiment.

Despite the mathematical complexities involved in calculating the average dimensions of real chains, it is possible to adopt a simplified model that lends itself to a very straightforward analysis. A model, in general, is a hypothetical representation of the real system. Although many of the details of the model are not realistic, we can derive a very good insight into the factors that influence the dimensional properties of a chain. Furthermore, it can be shown that for sufficiently high molecular weights many of the major features of a real chain will be similar to those of the model. We take as our model a series of n identical bonds, each of length l, which are joined together in a linear sequence. The simplifications take place as we remove the requirement that valence angles between successive bonds be maintained and also allow all rotational angles to be equally probable. This chain model thus has the property that the junction of any two bonds acts as if it were a universal joint. Figure 3.11 shows a two-dimensional representation of such a freely jointed chain which consists of 50 bonds. Each bond is represented by a vector of equal length. The position and orientation of the bonds are uncorrelated with one another. We seek the quantity $\langle r^2 \rangle$ for a chain containing n such bonds. To obtain our results, however, we must first find the probability $P(r)$ that the two ends of the chain are separated by the distance r.

The problem of obtaining the chain dimensions is identical to the classical mathematical problem of a random walk. In the simplest form of this problem the walker

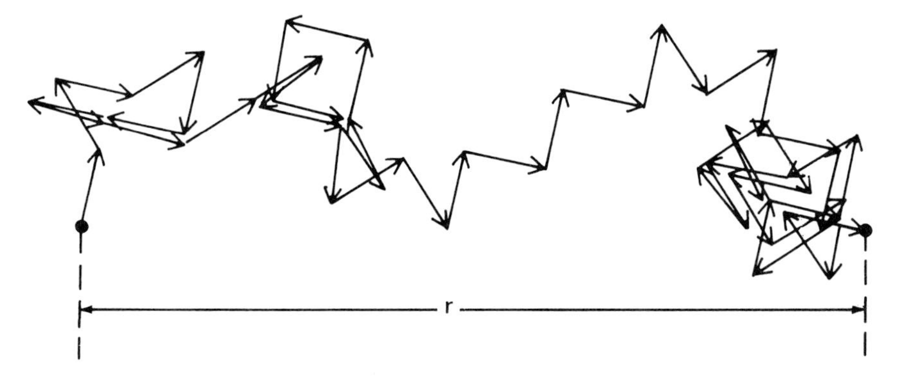

Figure 3.11. Two-dimensional representation of a freely-jointed chain consisting of 50 bonds. Reprinted from Paul J. Flory, *Principles of Polymer Chemistry*, Copyright 1953 by Cornell University. Used by permission of Cornell University Press.

takes a succession of steps of equal length which are uncorrelated with one another. In selecting the direction for a step, all memory of the path already traversed is erased. Thus, the freely jointed chain in two-dimensions, illustrated in Figure 3.11, can equally well represent a random walk. The starting point or origin is at one end of the molecule, while the terminus can be represented by the other end. However, even a freely jointed chain requires a three-dimensional representation. An analogous problem is seen in the random flight of a bird where the restriction of a planar path has been removed.

For either the random walk or random flight we require the distance traversed from the origin after n steps are taken. The probability that the path will be a straight line, which will require all the steps to be in the same direction, will clearly be exceedingly small. Similarly, the probability of the walker's returning to the starting point will also be very small. The exact mathematical solution to this problem, for all values of r, is given by the well-known error function of Gauss. The probability $P(r)$ is plotted graphically as a function of r in Figure 3.12. The corresponding mathematical expression is given by Eq. (1).

$$P(r) = Ar^2 \exp(-b^2r^2) \tag{1}$$

In this equation A and b are constants. The constant A is taken so as to assure us that the total probability (for all possible values of r) is equal to unity or certainty. The constant b is related to the properties of the freely jointed chain through the relation

$$b^2 = \frac{3}{2nl^2}\,. \tag{2}$$

The curve in Figure 3.12 has fulfilled our intuitive expectation. The probability of returning to the origin is zero. The probability of traversing very large distances is so small as to be negligible for all practical purposes. At some intermediate value of r, the probability has a maximum value. A random walk is therefore not a very efficient process by which to travel between two fixed points.

Our desired average quantity can be obtained from Eq. (1) by the standard method of averaging. It is found that

$$\langle r^2 \rangle = nl^2. \tag{3}$$

This equation tells us that the average of the square of the distance between the chain ends is proportional to the number of links in the chain. The proportionality constant is the square of the bond length. The appropriate linear dimension is the root-mean-square value of r and is expressed as

$$\langle r^2 \rangle^{1/2} = n^{1/2}\, l \tag{4}$$

From Eq. (4) we find that a linear dimension of the chain will increase only in proportion to the *square root* of the number of steps. This quantitative analysis of a random walk explains why a freely jointed chain will be much more compact than

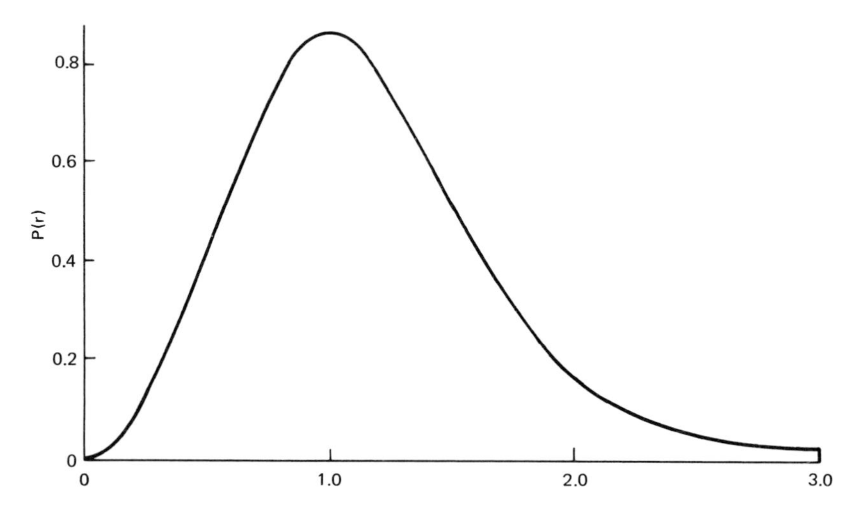

Figure 3.12. Curve representing the probability of finding the two ends of a freely-joined chain separated by a distance r. Graphical representation of Eq. (1). From L. R. G. Treloar, *The Physics of Rubber Elasticity*, The Clarendon Press, Oxford. Reproduced by permission.

an extended or stretched-out chain. In this latter conformation a linear dimension must of course be proportional to the number of bonds in the chain.

We have recognized earlier that a real chain molecule cannot be considered to be freely jointed. Therefore, it cannot be legitimately subjected to this particular random walk analysis. The influence of valence angle restrictions as well as the preference for certain rotational angles must be taken into account in describing the dimension of a real molecule. Fortunately, a detailed analysis shows that these factors can be properly and exactly taken into account merely by inserting a constant into Eq. (3). Thus, for a real chain we can write

$$\langle r^2 \rangle = Cnl^2. \tag{5}$$

The constant C will vary from one chain type to another and will depend on the chemical nature of the repeating unit, as is reflected in the bond angles and the hindrance to rotation. Units with different kinds of bonds can also be accounted for. Invariably, however, the proportionality between $\langle r^2 \rangle^{1/2}$ and $n^{1/2}$ is maintained for all real chains. For polyethylene, the constant C is found to be 6.8. Thus, for a chain where $n = 10^4$ and $l = 1.53$ Å, we find by substituting in Eq. (5) that $\langle r^2 \rangle^{1/2} = 320$ Å. There are several experimental methods that enable us to determine the quantity $\langle r^2 \rangle^{1/2}$ directly. Very good agreement is found between the theory and direct observation. On the other hand, a fully extended chain (all *trans* rotational angles) containing the same number of bonds will be 12,700 Å long. Thus, we can see quantitatively that the average end-to-end distance of a real chain, in a statistical conformation, is severely reduced relative to its extended length. If we take the value of $\langle r^2 \rangle^{1/2}$ to represent the diameter of the domain of the random coil, the volume for the above chain is 3.5×10^7 Å^3. This volume is very much greater than the volume actually occupied by the constituent atoms.[2]

The influence of the chemical nature of the repeating unit on the average dimensions of chains in statistical conformation is manifested in the value of the constant C. This very important constant can be calculated theoretically for a chain with a given repeating structure and can also be determined experimentally. A compilation of typical values for different polymers are given in Table 3.1. The values of this constant for real chains range from 2 to well over 100. To establish a point of reference, we note that for a freely rotating polyethylene chain C=2; the actual chain is somewhat more extended since it is found experimentally that C=6.8. However, this value of C still represents a relatively compact molecule according to Eq. (5). Most of the values listed in the table lie between 2 and 10. While from a very general point of view they can be considered to be compact, a discrimination can be made between the different values that reflects the specificity of the chemical repeating units and the concomitant interactions. For example, for the poly-α-amino acids C is about 8-10 when either straight chain or bulky side groups are attached to the α-carbon atoms. However, when the side group is replaced by hydrogen as in polyglycine, the molecule becomes much more compacted because of the reduced steric interraction. The reduced spatial extension is reflected in the fact that for this chain C is lowered to 2. For the poly imino acid, poly-L-proline, where the side group is a five-membered ring joined to the main chain, the molecule becomes more extended because of restrictions to bond rotations.

The *p*-phenylene type aromatic polyamides, as well as the corresponding polyesters, are very extended molecules, as is indicated by the constant C. This results because the ring structure is an integral part of the chain. The highly extended character of these chains endow them with unique molecular properties that lead to substances possessing unusual physical and mechanical properties (see Chapter 5). These type chains do not display random coil behavior except in the limit of exceptionally high molecular weights. On the other hand, chains characterized by smaller values of the constant C achieve random, or statistical, conformation at degrees by polymerization that are the order of 20-50. The degree of polymerization is defined as the number of repeating units in a molecule.

Ordered Conformations

Our discussion of chain conformations has been focused primarily on the disordered or statistical conformations. We have tactily assumed that successive rotational angles will be uncorrelated with one another. Molecular forms are thus gener-

[2] In the treatment of chain molecules within the framework of random flight theory, a fundamental difference between the trajectory of a bird and the conformation of a chain has been neglected. A bird in random flight can recross its path with impunity as many times as is desired. However, since a chain element has a finite volume, intersecting paths are not allowed since two elements cannot occupy the same space. An exact mathematical analysis that eliminates intersecting conformations has not as yet been accomplished. An approximate, but very reliable solution indicates a modification of the dependence on n in Eq. (3). This important result does not, however, impose any serious constraint for our purposes.

Table 3.1. Characteristic Dimensions of Chain Molecules*

Polymer	Repeating unit	Temperature °C	$C = \langle r^2 \rangle / nl^2$
Polyethylene	$-CH_2-CH_2-$	140	6.8
Polypropylene (isotactic)**	$CH_2-CH(CH_3)$	145	5.7
Polyisobutylene	$CH_2-C(CH_3)_2-$	24	6.6
Polymethyl methacrylate (isotactic)**	$CH_2-C(CH_3)(COOCH_3)-$	25	9.1
Polymethyl methacrylate (syndiotactic)**	$CH_2-C(CH_3)(COOCH_3)-$	35	7.2
Polyoxyethylene	$O-CH_2-CH_2$	36	4.0
Polydimethyl siloxane	$-O-SI(CH_3)_2-$	20	6.2
Polyhexamethylene adipamide	$-NH(CH_2)_6-NH-$ $-CO(CH_2)_4CO$	25	5.9
Poly α amino acids	$-NH-CH(CH_2R')-CO-$	25-100	8.5-9.5
Polyglycine	$-NH-CH(H)-CO$	25	2.0
Poly-L-proline	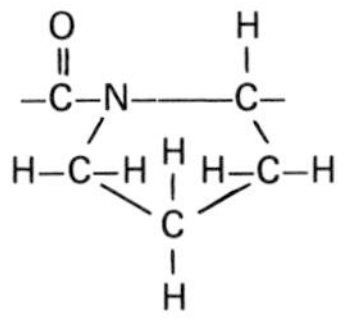 $-C(=O)-N-C(H)-$ (ring: $N-CH_2-CH_2-CH_2-C$)	30	13-20

Table 3.1. *(Continued)*

Polymer	Repeating unit	Temperature °C	$C = \langle r^2 \rangle / nl^2$
Natural rubber	$CH_2{-}C(CH_3){=}CH{-}CH_2$	50	4.7
p-Phenylene polyamide	$-N(H)-C_6H_4-N(H)-C(=O)-C_6H_4-C(=O)-$	30	125

*Adapted in part from P. J. Flory, Statistical Mechanics of Chain Molecules, Wiley (1969).

**The two sterioisomers of vinyl polymers will be discussed in the next section.

ated that do not represent any well-defined shapes. These structures, typical of all chain molecules, represent an overwhelming preponderance of the available conformations. They indeed serve as the basis for a large number of properties. However, we must recognize that among these multitudinous conformations there exist a very few distinct ones in which successive rotational angles are correlated with one another. A definite set of rotational angles will be repeated in a very prescribed manner along the chain. We have already encountered one such simple example in the extended planar zigzag form of polyethylene. Here each rotational angle assumes the *trans* position. More generally, when a given sequence of rotational angles successively repeats itself, an ordered structure will result and the molecular shape will have a rod-like character. In analogy, while a random walk leads to a disordered conformation, a directed or defined walk results in an ordered conformation. In non-molecular terms, a comparison can be made with a long piece of spaghetti. Before being cooked, it has the form of a long thin rod. After cooking, a much more compacted, coiled up, irregular structure results, which is continually changing its size and shape. The former situation can be thought of as representing an ordered structure. The latter, a statistical or random one.

Ordered structures of isolated chain molecules can be observed in very dilute solutions under certain circumstances. A regular chain conformation is maintained that requires either very specific intramolecular bonding or strong interactions. The required regularity can also be maintained by having rather severe and repetitive steric repulsions. Ordered chains developed by specific interactions are found in macromolecules of biological interest such as the polypeptides in α-helical forms and the native and synthetic nucleic acids. These classes of macromolecules are examined separately in Chapter 6. Examples of ordered structures stabilized by steric interactions are not as common. They are, however, extremely important and a great deal of effort is going into the synthesis and development of such chains. An example of molecules in this class is given by the poly (N-alkyl isocyanates) such as

$$-C(=O)-N(R)-C(=O)-N(R)-$$

Here, the highly restricted rotation about the C-N bond and the steric repulsion between substituents attached to the nitrogen atom allows for a highly ordered form to be maintained in an isolated molecule. The *p*-phenylene polyamides and polyesters show similar characteristics because of the rotational restrictions imposed by the chain ring. For either of the causes cited above, the long chains will adopt a rod-like form that is in marked contrast with the random coil structure that is typified in Figure 3.9. Because of their high molecular weight, these molecules are geometrically very asymmetric and can be envisioned as long, thin rods. Their length will be many orders of magnitude greater than their breadth. Lengths that are of the order of hundreds to thousands of Angstroms and diameters in the tens of Angstroms can be easily developed.

The dispersion of such rod-like particles of high axial ratio in a liquid medium presents some unique problems with important ramifications. The shape anisotropy of the particles presents space-filling problems in a fixed volume. The molecules can be randomly arranged relative to one another only at high dilution. Put another way, an isotropic, or uniform, solution of rod-like molecules, which is schematically illustrated in Figure 3.13, can only exist in very dilute solution. For geometric

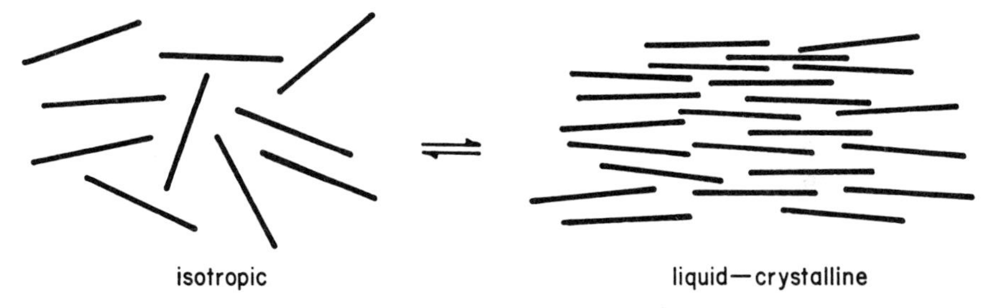

Figure 3.13. Schematic representation of isotropic and liquid-crystalline solutions of highly asymmetric rod-like molecules.

reasons, there is simply not enough room to accommodate many such molecules in random disposition. As the polymer concentration is increased then because of the space-filling requirement, a random arrangement can no longer be maintained. In order for the solution to continue to exist the particles must become preferentially aligned with respect to a common axis. In this way the stringency of the space-filling requirement is overcome. Thus, as is shown in Figure 3.13, a liquid-crystalline phase, which is anisotropic, develops. This anisotropic phase, which can form solely on the basis of the molecular shape and can be further enhanced by intermolecular repulsions, is only one-and-a-half to two times more concentrated than the dilute isotropic phase. As we shall discuss in more detail in Chapter 5, this process by which preferential molecular orientation can be achieved has important implications in the development of a new class of synthetic fibers and could very well be playing an important role in fibril formation in natural occurring systems.

For most of the chain molecules, which do not adhere to the above conformational strictures, ordered structures are observed only under special conditions of temperature and pressure when a large number of them are brought together in the solid or crystalline state. Under these circumstances, it is possible to discuss the conformations or structure of a single chain. To form a stable, ordered chain structure the set of rotational angles that are successively repeated must be such that

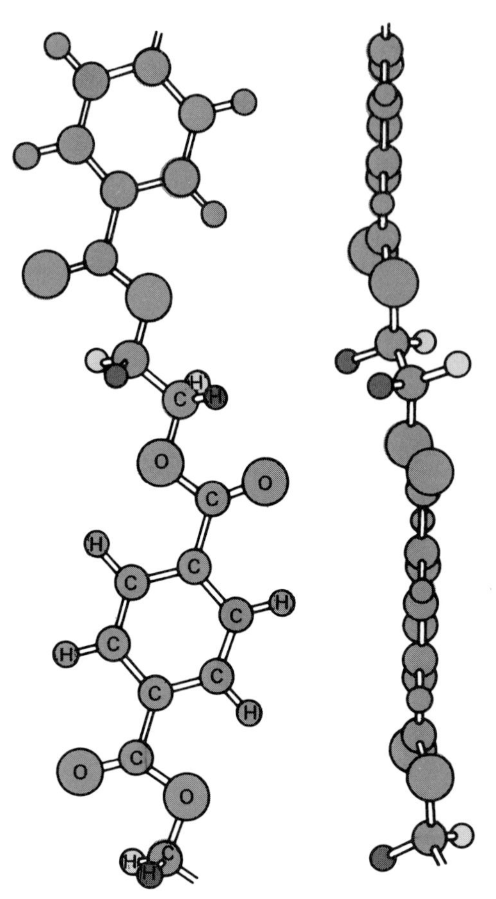

Figure 3.14. Ordered chain conformation of polyethylene terephthalate. From R. Dubeny, C. W. Bunn and C. J. Brown, Proceeding Royal Society *226A*, 531 (1954). Reproduced by permission.

each angle represents an energy minimum. Thus, a very energetically favored conformation is developed. For polyethylene, this condition is achieved when each bond assumes the *trans* position, as was illustrated in Figure 3.1.

A planar, or nearly planar, extended ordered conformation is characteristic of many polymers, such as the polyamides and polyesters. For example, as is shown in Figure 3.14, the observed ordered structure of the aromatic polyester, polyethylene terephthalate, is nearly planar. The chain length is just slightly less than that for full extension. For aromatic polyamides, in ordered conformation, the chains are in a planar zigzag conformation and are fully extended. The ordered structures of two commonly used polyamides, nylon 6-6 and nylon 6, are represented in Figure 3.15. For each of these polymers, three chains arranged side by side are shown. The planar zigzag form is readily apparent for each of the individual chains. In this conformation, the geometry and atomic coordinates are such that a very favorable strong interaction takes place between the C=O group of one chain and the N=H group of the adjacent chain. Such interactions are known as hydrogen bonds. Every possible

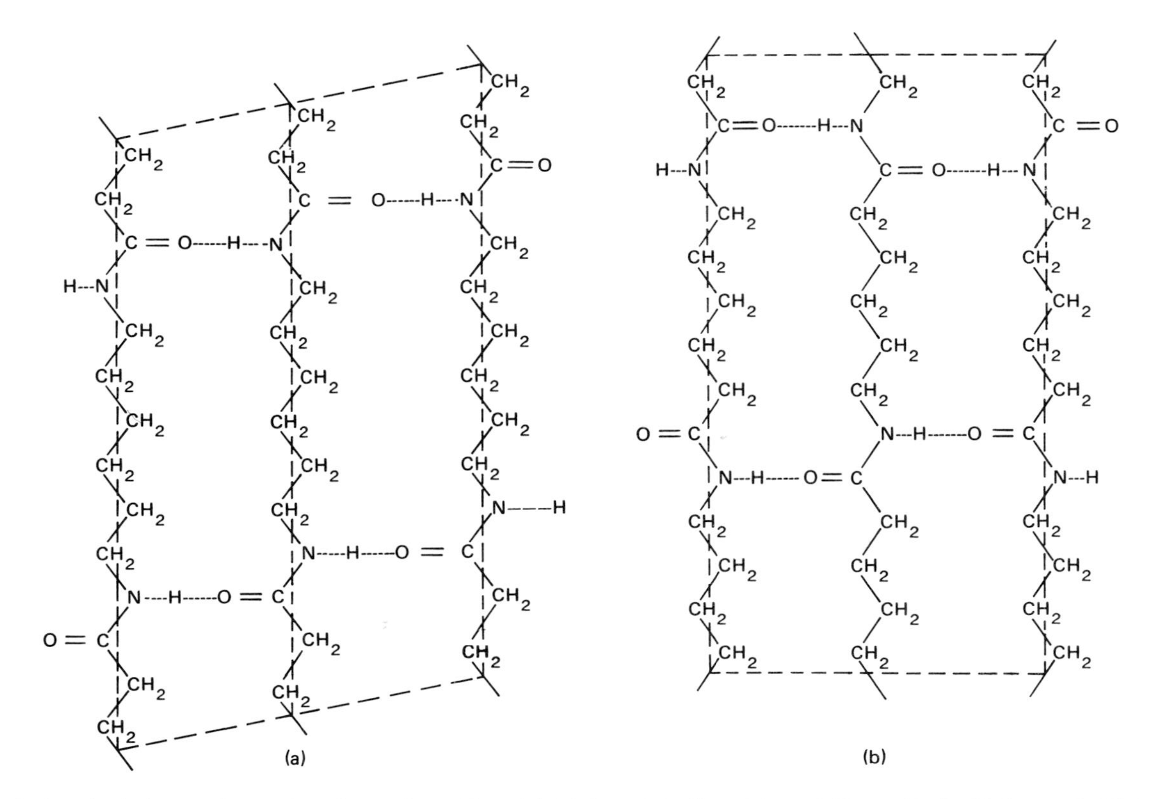

Figure 3.15. Ordered chain conformation for polyamides. (a) Nylon 6-6; (b) Nylon 6. From D. R. Holmes, C. W. Bunn and D. J. Smith, Journal of Polymer Science *17*, 159 (1955).

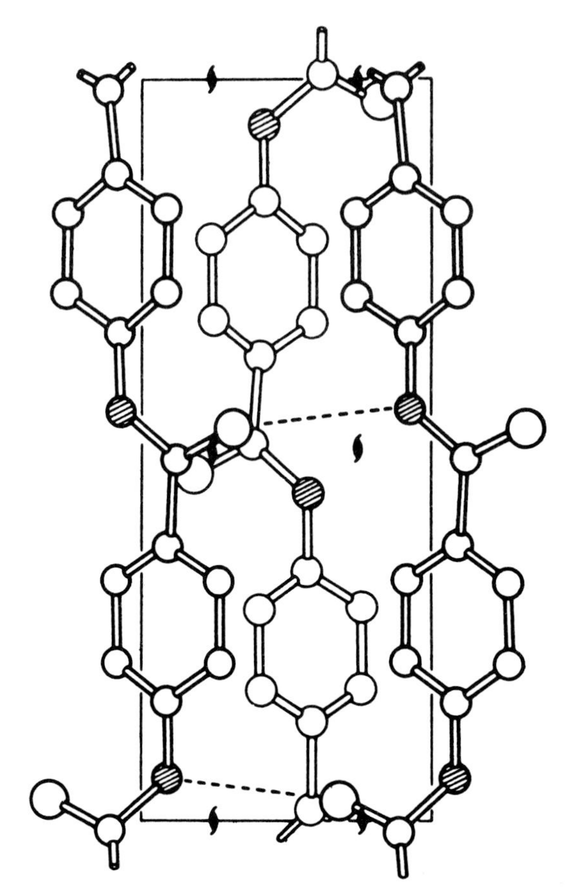

Figure 3.16. Ordered structure of *p*-phenylene polyamide. From H. Tadokoro, *Structure of Crystalline Polymers*. John Wiley & Sons, New York. Reproduced by permission.

C=O–H–N hydrogen bond is formed for these chains. These exert a very strong influence on the physical properties of the nylons.

The ordered structure of a *p*-phenyl polyamide is given in Figure 3.16. It is apparent from this diagram how the presence of the ring in this chain leads to an extended structure. In the crystal lattice of this polymer (as distinct from the individual isolated molecules), the chains form hydrogen bonded sheets that are very similar to those found in the aliphatic nylons. However, the unusual strength and rigidity of this molecule results primarily from the ring being part of the chain backbone and its influence on conformation and subsequent orientation.

When one or more of the hydrogens from a carbon-carbon single bonded chain are replaced by other atoms or groups of atoms, the planarity of the structure is very often disturbed. Substituents larger than hydrogen can cause severe steric or packing problems when the *trans* orientation of each bond is maintained. To alleviate this very unfavorable energetic situation, other rotational angles, or succession of rotational angles, must be adopted. When these deviations from the *trans* position are repeated in a systematic manner, a spiral or helical form of the chain must evolve. We can illustrate this point in a purely formal way by allowing a *trans-gauche* sequence to be perpetuated along the chain. As is diagrammatically represented in Figure 3.17, a helical structure results. Since there are two possible *gauche* positions obtained by rotation of plus or minus 120° from the *trans* (see Figure 3.6) the same *gauche* angle must always be taken to preserve the order. Depending upon which of the *gauche* orientations is selected, either a right- or left-handed helix will be gener-

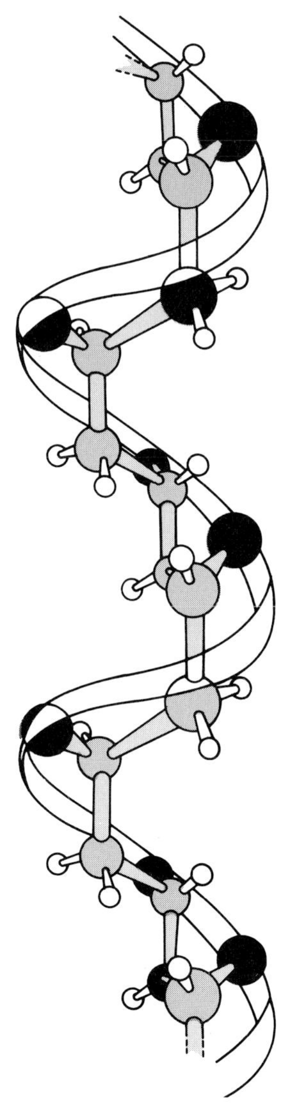

Figure 3.17. Diagrammatic representation of helical structure that is generated by a succession of *trans*-gauche rotational angles. Backbone carbon atoms, gray; substituent groups, black; hydrogen atoms, white.

ated. The planar zigzag form is merely a special case of a fully stretched out spiral which is restricted to lie in a plane.

A simple example of the deviations from planarity is seen in the case of polytetrafluoroethylene, or Teflon. In this molecule all the hydrogen atoms of polyethylene have been replaced by fluorine. Because of the increased atomic size, albeit a very modest one, the all *trans* structure can no longer be maintained. As is shown in Figure 3.18, the chains have to be twisted out of the plane at each carbon atom by 14° to accommodate the fluorines. A spiral structure results, which repeats itself after every 13 carbon atoms. As is indicated in the figure, carbon atoms labeled 1 and 14 are thus in equivalent spatial positions.

Another molecule of interest is polyisobutylene. Here the pairs of CH_3 groups

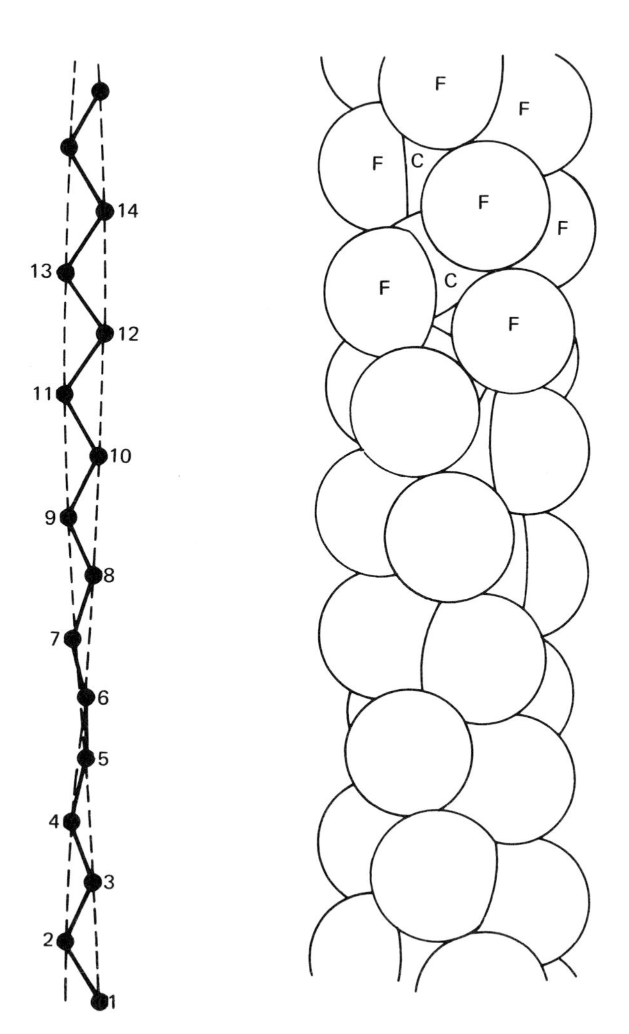

Figure 3.18. Ordered, twisted conformation of polytetrafluoroethylene. From C. W. Bunn and E. R. Howells. *Nature 174*, 549 (1954). Reproduced by permission.

$$\left(\begin{array}{cc} H & CH_3 \\ | & | \\ -C & -C- \\ | & | \\ H & CH_3 \end{array} \right)_n$$

on the alternate chain carbon atoms cause a very severe overcrowding. These steric difficulties cannot be overcome by any combinations of bond rotations, which are restricted to the *trans* or *gauche* states. A stable, ordered helical structure can be formed with this chain by rotating each bond 82° from its *trans* state. This rotation generates a helix in which eight chemical repeating units correspond to five turns of the geometric repeating unit. An ordered structure will result when the direction of the rotation is the same for each bond. However, when the direction of the rotation is allowed to change at alternate bonds, a disordered statistical conformation will result. The ordered conformation of polyisobutylene was one of the first helical structures discovered in long-chain molecules.

Vinyl polymers of the type $\left(\begin{array}{cc} H & H \\ | & | \\ -C & -C- \\ | & | \\ H & R \end{array} \right)_n$, in which dissimilar substituents are attached to alternate carbon atoms, present a very special situation, since two different structures are possible. This situation is illustrated in Figure 3.19 for a repeating unit selected from within the chain. In this figure the carbon-carbon bond, which forms the chain backbone, lies in the plane of the paper. Attached to one of the carbons are the two hydrogen atoms, one directed above the plane (solid line) and the other directed below (dashed line). These two hydrogens are equivalent to one another and are thus the same for each of the structures. However, for the other carbon atom in which two chemically different substituents are attached, the situation is not the same. In Figure 3.19(a) the CH_3 group lies above the plane, while the hydrogen atom is below. In Figure 3.19(b), however, the opposite situation exists—the hydrogen atom is directed above the plane, while the CH_3 is directed

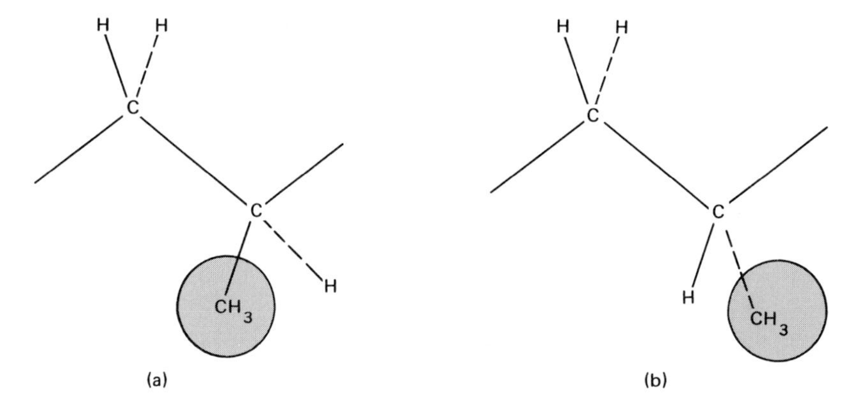

Figure 3.19. Alternate arrangements of substituents about carbon-carbon bond of polypropylene.

below. Neither forms can be converted into the other by any possible internal rotations about the C–C bonds or any symmetry operation involving the whole chain. They are distinctly different chemical structures called stereo-isomers, which are mirror images of one another. They are related to each other in the same way as the left hand is to the right. A chain comprised of both of these units is a stereo-irregular copolymer, and is given the name atactic.[3] When only one of the stereo-isomers is perpetuated along the chain, a stereo-regular polymer called isotactic is formed. Stereo-regular polymers are also formed when there is a regular alternation between the two isomers along the chain. Such a chain has been named syndiotactic. Chains with a high degree of stereo-regularity were first synthesized in the 1950s by Professor G. Natta of Milan. The polymers were prepared by ionic addition polymerization methods using a stereo-specific catalyst developed by Professor K. Ziegler of Germany. For this outstanding accomplishment in synthetic chemistry, Natta and Ziegler received the Nobel Prize in 1963.

The stereo-regular forms of such chains can be schematically represented as in Figure 3.20(a) and (b). The atactic form is shown in Figure 3.20(c). We take the specific case of polypropylene, R = CH_3, and to illustrate the stereo-isomers, take the carbon-carbon backbone to be planar zigzag. In the isotactic structure all the CH_3 groups are positioned above the plane (or all below); while in the syndiotactic form there is a regular alternation of the CH_3 groups above and below the plane. In the atactic form the CH_3 groups are randomly positioned along the chain. If, however, we examine the planar zigzag form of isotactic polypropylene more closely, we find the distance between successive CH_3 groups is only 2.5 to 2.6 Å. This small separation cannot be tolerated, since it leads to gross overcrowding of the methyl groups. This severe steric problem can be overcome very simply by having alternate chain bonds assume a *gauche* orientation. As we have discussed, the planarity of the chain is now lost and a simple helical form evolves. In this new geometrical pattern the CH_3 groups are now 3.2 Å apart. This is a sufficient separation to eliminate the unfavorable steric interaction. As is shown in Figure 3.21, the helical structure particular to polypropylene contains three chemical repeating units for each repeat of the geometric structure. Isotactic polymers, which contain bulkier side groups, will naturally require more space. Hence helices with larger repeats will be generated. For syndiotactic polymers the substituents will be further apart, as the diagram of Figure 3.20(b) indicates. The steric problem between neighbors is not as severe as for isotactic polymers. Planar zigzag ordered structures are therefore still possible and are in fact observed in many cases. An example of this is found in syndiotactic polyvinyl chloride.

In this chapter we have summarized the conformational or structural properties of long-chain molecules. There are two major conformational categories. The disordered or statistical conformations encompass most of the allowed possibilities. However, there also exist very important and unique highly ordered chain structures. There is a very distinct similarity and analogy here between the liquid and crystalline (solid) states of monomeric substances. The liquid state is characterized

[3] Tactic comes from the Greek word meaning order.

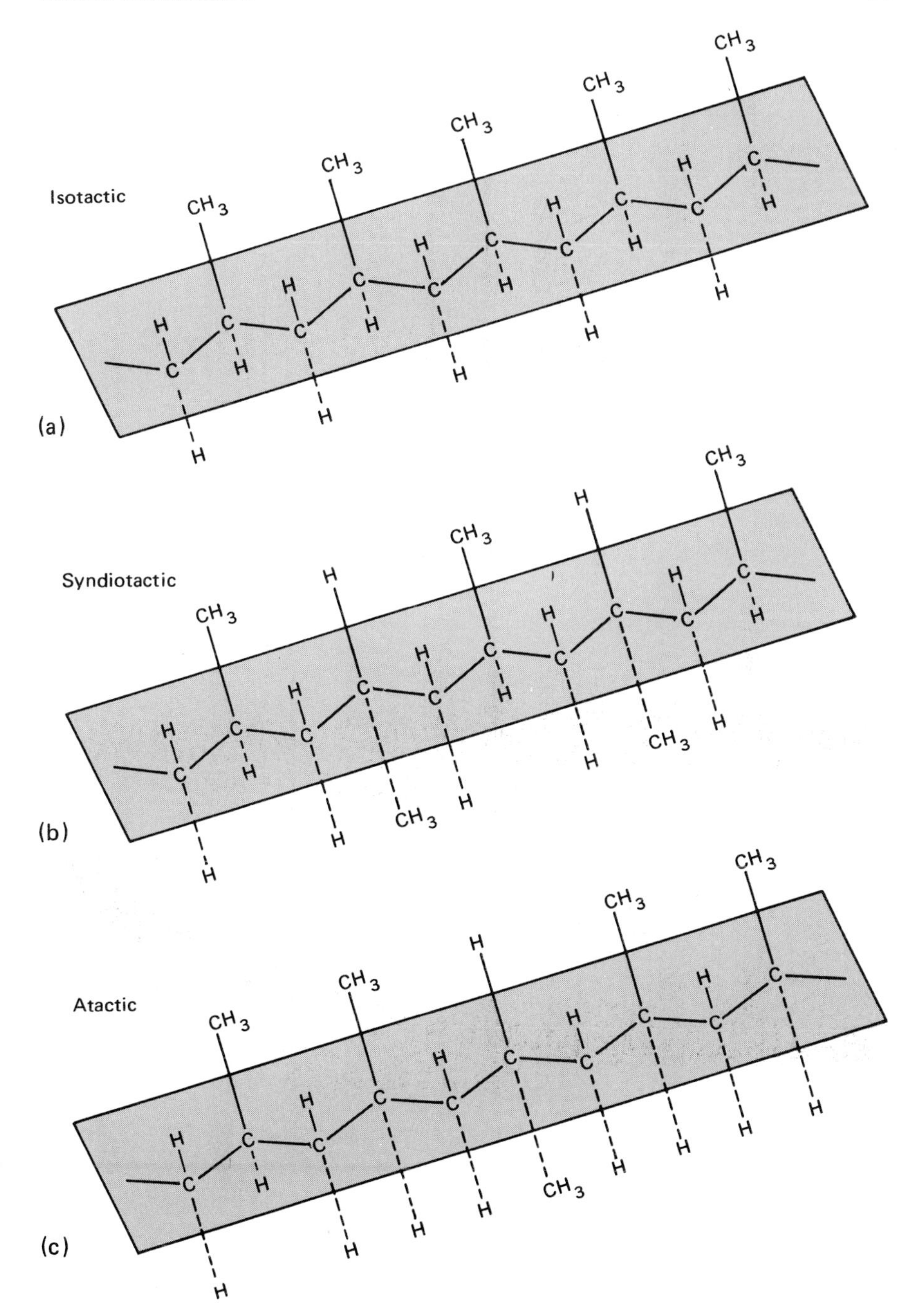

Figure 3.20. Schematic representation of stereo-isomeric forms of polypropylene.

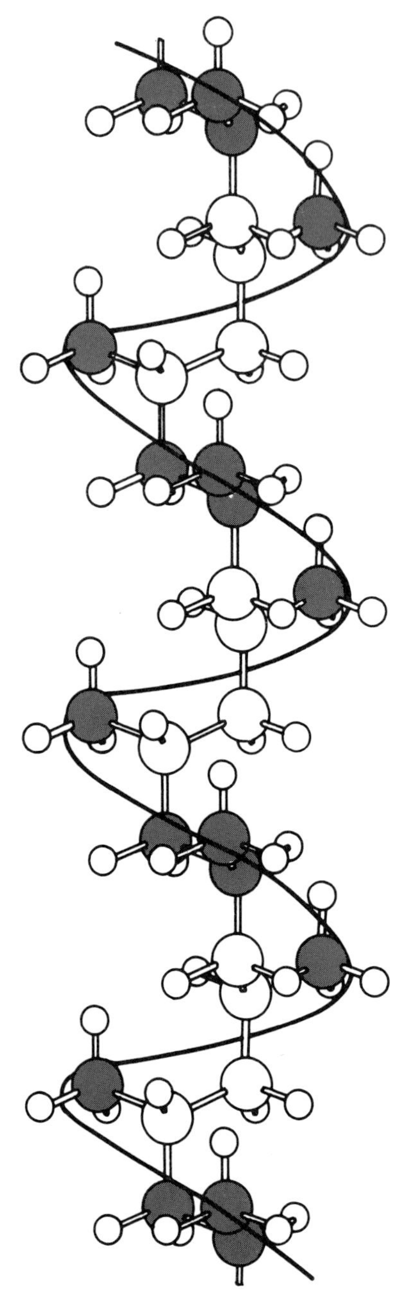

Figure 3.21. Ordered helical conformation of isotactic polypropylene. Adapted from F. Danusso, *Polymer 8*, 281 (1967).

by the disorderly, random arrangement of the molecules. The crystalline state is represented by a precise and orderly three-dimensional molecular arrangement. We have also described the special and important situation of highly asymmetric rod-like macromolecules forming liquid-crystal phases that are comparable to those formed by low molecular weight substances. We should, therefore, expect than when a collection of macromolecules is brought together to form a macroscopic or observable system, the resultant properties will depend on the conformation of the individual chains. Put another way, the chemical and physical properties will depend upon whether or not the chains are in disordered conformations. In the following chapters, we examine in more detail the major properties of polymeric systems and their relation to chain structure.

Chapter 4

RUBBERS AND GLASSES

Rubbers–Introduction

The unusual elastic and mechanical properties of rubber and rubber-like materials are well known to us from everyday experience. Rubbers represent a unique class of substances since they can undergo very large deformations without breaking and spontaneously return to their original dimensions when the applied force is removed. A typical rubber can be reversibly extended from five to ten times its original length. Such a deformation is called an elastic one. In contrast ordinary monomeric solids, such as metals and crystals, can only be elastically deformed about 1 percent. Further deformation of these materials results either in rupture or in plastic flow. Plastic flow is an inelastic deformation, for when the applied force is removed the sample remains in the deformed state. Thus, the extensibility of a rubber is about 1000 times greater than an ordinary solid. Besides the differences in extensibility, the forces required to achieve the same deformation for these two classes of substances are very different. For example, if a steel wire and rubber filament of the same diameter were stretched only 1 percent of their original length a 10^5 times greater force would be required for the steel wire. In Figure 4.1 we compare typical force-length curves of rubbers and monomeric solids. The monomeric material is relatively inelastic since a large force is required to achieve a small extension. On the other hand, five- to sixfold increases in length are easily achieved with the rubber by the application of only modest external forces.

Long-range elasticity is a property found only in polymeric substances. However, it is not restricted to chain molecules with a particular kind of chemical repeating unit commonly associated with rubbers or elastomers. It is in fact observed for all types of long-chain molecules, irrespective of their origin, when they exist in statistical conformation, i.e., in the disordered state. Under appropriate conditions of temperature and composition, identical long-range elastic properties are observed in polyesters and polyamides in denatured globular and fibrous proteins, and in the protein elastin found in elastic tissue and in the walls of blood vessels. We must conclude from the diversity of chemical types that display this elastic behavior that it is the result of a common underlying structural feature unique to long-chain molecules. It is a well-established generalization that synthetic macromolecules, as well as those of natural origin, display similar characteristics in the disordered state.

The unique elastic properties of rubber-like materials were first brought to attention by J. Gough in 1806. Gough wrote that "the substance called Cautchouc or Indian Rubber possesses a singular property which, I believe has never been taken notice of in print.... so that when a slip of this resin has been sufficiently warmed, it may be extended to more than twice its natural length, by a moderate force applied to its extremities, after which it will recover its original dimension in a moment, provided one of the ends of it be let go as soon as it has been stretched." In this classical communication Gough pointed out two extremely important ther-

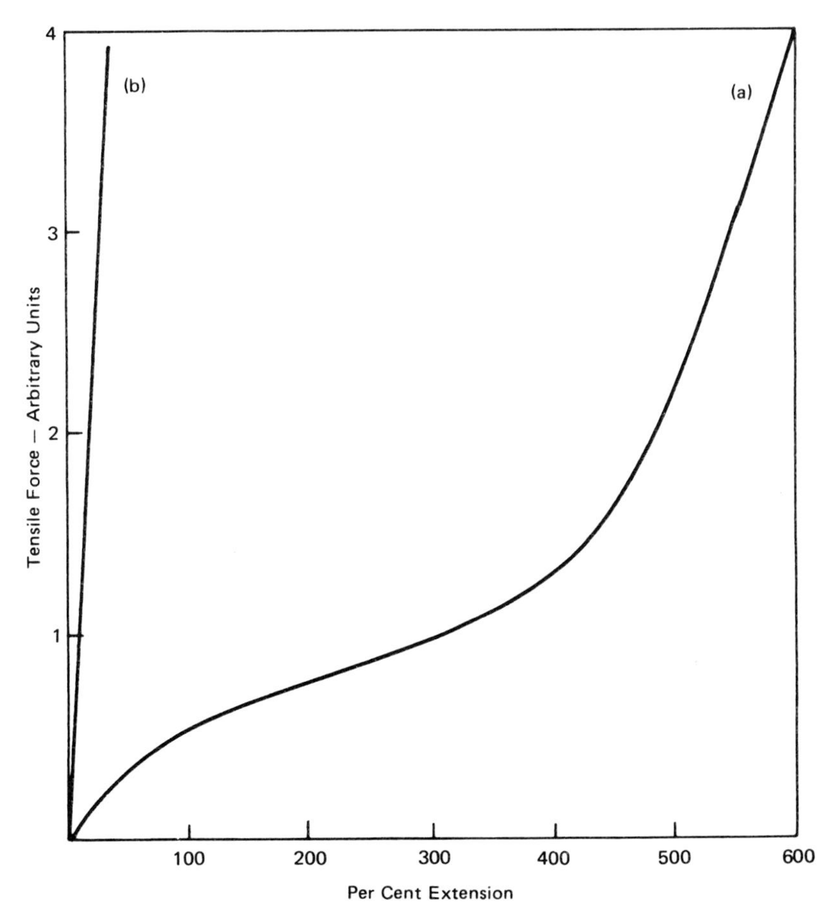

Figure 4.1. Schematic representation of force-extension curve for (a) a rubber-like substance; (b) a monomeric solid.

mal and thermoelastic properties of rubber. It was well known that the usual or more common substances expand on heating, whether they are in an unstressed or stressed condition. This normal thermal expansion behavior is obeyed by virtually all monomeric substances.[1] Polymeric materials, when undeformed, follow this classical pattern. However, Gough discovered almost two centuries ago that when a stretched piece of rubber is heated, its length diminishes. These two kinds of thermal expansion are illustrated in Figure 4.2. Stretched rubbers thus have a negative thermal expansion coefficient; contraction is observed in heating and expansion on cooling. This unique behavior is found in all deformed polymeric systems when in the disordered state.

Closely related to the negative thermal expansion coefficient was his observation that when a rubber sample is extended rapidly its temperature rises. This phenomenon can be demonstrated very simply by rapidly extending a rubber band and imme-

[1] An obvious exception to this generalization is the behavior of water at 4°C.

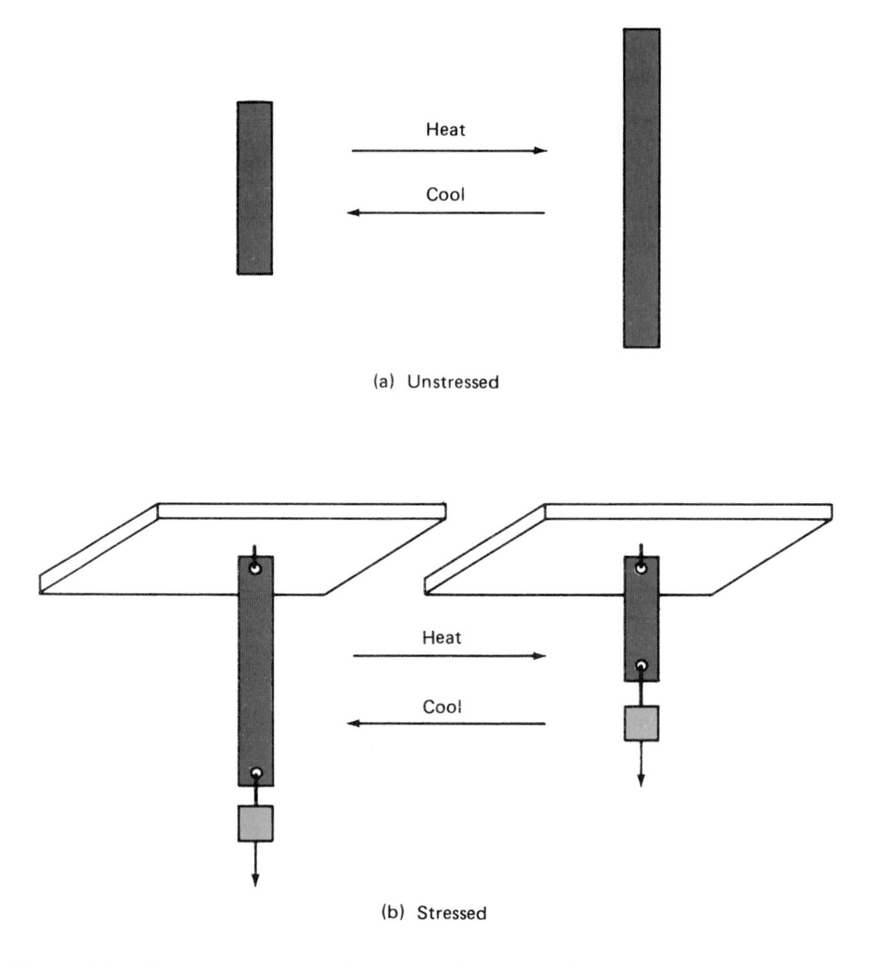

Figure 4.2. Thermal expansion behavior of a rubber-like substance (a) unstressed; (b) stressed.

diately placing it on one's lips. The change in temperature is easily detected. More sophisticated measurements show that there is about a 10°C temperature rise for a four- to fivefold extension. A student of thermodynamics will recognize that these rather special thermal properties are directly related to one another. From the Second Law of Thermodynamics one can conclude that the heat evolved during the deformation of a rubber-like substance requires an accompanying decrease in the entropy. The observed thermal-elastic behavior is in turn a direct consequence of this entropy decrease.

The characteristic molecular structure of polymers in the disordered state allows for an understanding of the underlying basis for the long-range elasticity and for a direct connection to be made between the unusual mechanical and thermal properties that are observed. Macromolecules in the unstrained state possess a tremendous conformational versatility as a result of the rotational freedom about the single bonds comprising the chain. A variety of structures, ranging from very compact to more extended ones, are thus available to the system. When subject to an external force, these conformations will be rearranged. In particular, the more extended forms will now be the ones that will be favored. A polymeric substance can, therefore, sustain large deformations without rupture because molecular structures are available that are consistent with the increase in macroscopic dimensions. The deformation capability is thus a reflection of the change in molecular arrangement without requiring any change in the intrinsic molecular structure. When deformed, the total number of conformations available must, however, be reduced. A decrease in the entropy results, as is demonstrated by the thermal properties that are observed.

Since the original, undeformed dimensions of a rubber-like material can be easily regained, a large restoring force must have developed. This restoring force is a reflection of the strong desire of the chains to return to their original statistical conformation. Put another way, the force causing a rubber-like substance to snap back is merely the manifestation of the desire to return to the most probable state. In this state the system can assume all possible conformations, unbiased by the external stress. The most probable state, that of the maximum entropy, is desired by all molecular systems. This particular property is not special to polymeric substances. Thus, the unique property of long-range elasticity is directly and intimately associated with the characteristic and novel conformational properties of polymers. In contrast, the elastic deformation of monomeric solids and liquids results primarily from the displacement of individual atoms relative to one another. The molecular mechanism for these substances is quite different from that for polymers. It is, there, not surprising that for monomeric materials the maximum deformation that can be attained and the thermal properties are also different.

The fact that rubber elasticity is based on simple, well-established physical laws can also be seen in the conceptual similarity that exists with the compression of an ideal gas. For example, when the volume occupied by a gas is reduced by increasing the pressure, then the number of configurations or arrangements of the gas molecules is also reduced. The entropy of the system is decreased, and heat is evolved as expected. As we have already noted, when a rubber band is stretched, the entropy is also decreased and heat is evolved. The pressure exerted by an ideal gas, at constant volume, is directly proportional to the absolute temperature. We shall find that the force exerted by a rubber band, held at fixed length, is also directly proportional to the absolute temperature.

Structural Basis for Rubber Elasticity

The molecular concepts underlying long-range elasticity can also be given a more quantitative description. The formal methods of statistical mechanics are used to express the relationship between the number of conformations available to the system, the state of deformation and the change in the entropy. By following these procedures, the force-length relationship can be derived. As a first step in this process, it is instructive to develop an expression for the tension exerted on a single isolated long-chain molecule. This is clearly a hypothetical situation, since one cannot grab the two ends of an isolated molecule and move them apart. However, although we will ultimately be concerned with the macroscopic properties of a collection of such molecules, an understanding of the elastic properties of one such molecule provides the basis for understanding the macroscopic system.

For present purposes we shall take a freely jointed chain to represent our real molecule. The mean-square end-to-end distance, $\langle r^2 \rangle^{1/2}$, is then given by Eq. 4 of Chapter 3. The probability of finding a distance r between the two ends of the chain is given by Eq. 1 of Chapter 3 and is illustrated in Figure 3.12. We now imagine such a molecule with its ends separated by a distance r and held fixed at points A and B as in Figure 4.3. If the ends were not fixed, then the end-to-end-distance would be continually changing as the molecular conformation rearranges as a result of the thermal motion of the chain elements. When the chain ends are fixed, as in the deformed condition, the motion of the chain elements acts to reduce the distance r to its most probable value. This action generates a force directed along the line AB. Therefore, to maintain r at a fixed value, an oppositely directed external force must be applied to the chain ends. This force is indicated by the symbol f in the figure. The origin of this tensile force is thus analogous to the pressure developed on walls of a container by a gas molecule constrained to a fixed volume. By applying statistical principles it is found that the tensile force can be expressed as

$$f = 2kTb^2r \tag{1}$$

where k is Boltzmann's constant, T the absolute temperature, and b is the same constant that appeared in Eq. 1 of Chapter 3. We find, therefore, that the force acting along the line joining the ends of this isolated chain is directly proportional to the absolute temperature. It acts in the direction of the line joining the chain ends and is directly proportional to the distance between them. Thus, a deformed, isolated chain molecule obeys Hooke's Law, since the force is directly proportional to the displacement.

It now remains for us to relate the elastic behavior of an individual chain to that

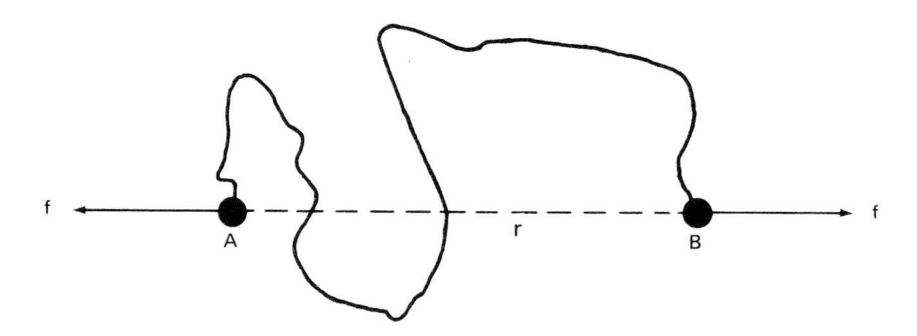

Figure 4.3. The force on an isolated chain whose ends are held fixed. Adapted from L. R. G. Treloar, *The Physics of Rubber Elasticity,* The Clarendon Press, Oxford. Reproduced by permission.

of a real macroscopic sample. Such a sample consists of a collection of a large number of chain molecules. We must first of all make sure that upon deformation the molecules will not slip or flow past one another. The flow will, of course, be much less than would occur in a monomeric liquid because of the entanglement and interweaving of the long chains with one another. However, it is still of sufficient intensity so that a pure elastic deformation will not result. Fortunately, this problem can be easily solved by introducing intermolecular cross-links to form a three-dimensional network structure. The number of cross-links required is not very large: about one-tenth of 1 percent of the chain repeating units is usually sufficient. This number is adequate to prevent flow, but sufficently small so as not to retard the segmental motion necessary for the conformational versatility. In rubber technology this cross-linking process is known as vulcanization. Charles Goodyear's discovery, in 1839, that natural rubber can be cross-linked with sulfur when the mixture is heated paved the way for the practical and commercial use of rubber. Although the fine details of the chemistry involved in Goodyear's vulcanization process are still not completely understood, the double bonds of the isoprene repeating unit are clearly involved. Schematically, the intermolecular cross-linking between two isoprene chains can be represented as shown in Figure 4.4. For this type of reaction,

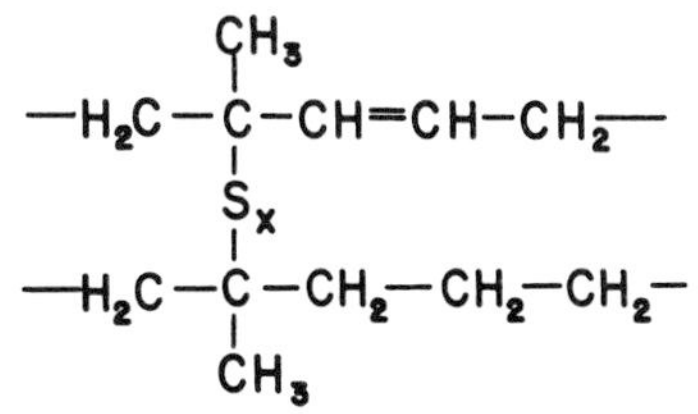

$$\begin{array}{c} CH_3 \\ | \\ -H_2C-C-CH{=}CH-CH_2- \\ | \\ S_x \\ | \\ -H_2C-C-CH_2-CH_2-CH_2- \\ | \\ CH_3 \end{array}$$

Figure 4.4. Schematic representation of natural rubber chains cross-linked with sulfur.

where units from different chains are joined in pairs, four chains will emanate from each cross-link in the network. This type of cross-linking is tetrafunctional. When a critical, predetermined, number of such cross-links are introduced into a collection of linear chains, a three-dimensional network, as was illustrated in Figure 3.6, is formed. When natural rubber is excessively cross-linked, however, its elastic properties are lost. It becomes a hard, rigid material known as ebonite.

A schematic representation of network formation followed by deformation is given in Figure 4.5. One starts with a collection of undeformed randomly coiled chains (Figure 4.5a). Intermolecular cross-links are than introduced to form a three-dimensional network. A chain can now be defined as that portion of the structure that extends from one cross-linkage to a successive one along the contour of a given molecule. Thus, for example, the sequence of units traversing the contour from A to B in Figure 4.5(b) represents one chain, whose vectorial end-to-end distance is represented by $\bar{r}$. Our network will be characterized by ν such chains, ν being the number of cross-linked units. Upon stretching, as is shown in Figure 4.5(c) the network becomes distorted. The chain contours naturally become more extended. The set of vectors that characterizes the end-to-end distances of the chains is considerably altered in both direction and magnitude. The vectors will become strongly

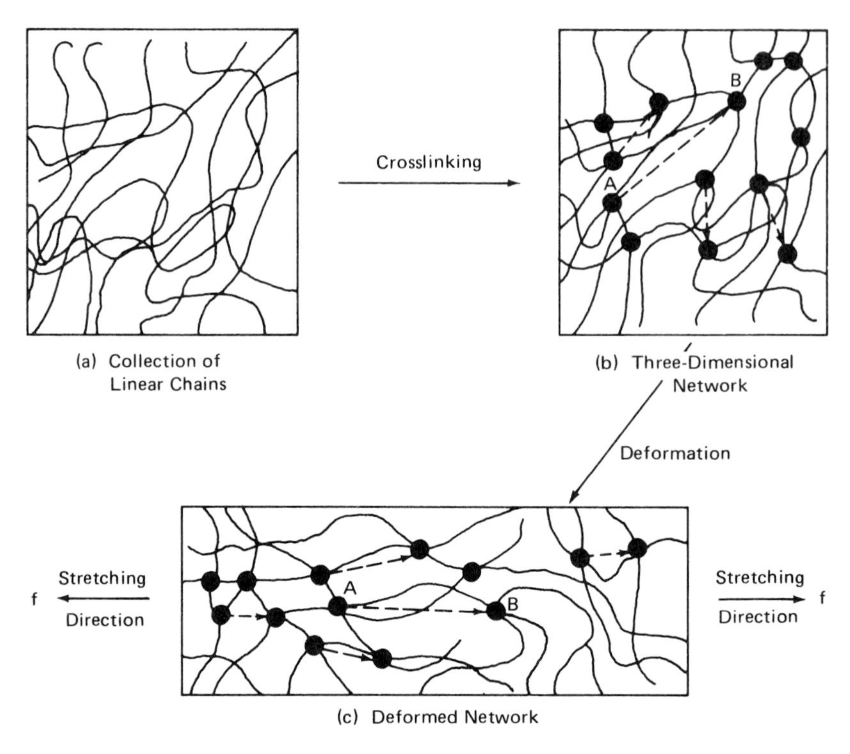

Figure 4.5. Schematic representation of network formation and deformation. Reprinted from Paul J. Flory, *Principles of Polymer Chemistry,* Copyright 1953 by Cornell University. Used by permission of Cornell University Press.

biased in the stretching direction, as can be seen from following the trajectory of the set of typical vectors from Figure 4.5(b) to 4.5(c).

Our problem now is to relate the macroscopic dimensions of the deformed sample to those of the individual chains making up the network. This is done by recognizing that the cross-linkages represent fixed points in the network structure. However, when a network is deformed, these junction points must take on a new set of average spatial positions relative to one another because of the distortions imposed by the emanating chains. The coordinates describing their position in space must be altered (relative to the undeformed state) in proportion to the dimensional changes of the bulk sample. Put another way, the alteration in the length and direction of the chain vectors must correspond to the macroscopic changes. This requirement is a natural consequence of the isotropic character of the network.

When the foregoing connection is made between the microscopic (molecular) and macroscopic dimensional properties, it is found that for simple extension, the stress corresponding to a length L can be expressed as

$$\tau = NkT\left[\frac{L}{L_0} - \left(\frac{L_0}{L}\right)^2\right] \tag{2}$$

where τ is the force per initial cross-sectional area, L_0 is the initial length, and N is the number of chains per unit volume (ν/V).[2] The theoretical result embodied in Eq. (2) points out the interesting fact that Hooke's law will not be obeyed by a bulk rubber sample. As will be recalled, this is contrary to the results obtained for an isolated chain.

In an exhaustive set of experiments involving a variety of rubbers of different chemical types, the force at fixed length is found to be directly proportional to the absolute temperature. This is one of the important conclusions drawn from Eq. (2). The factor in front of the brackets in this equation is the elastic modulus, which can be directly predicted from molecular theory. Within the uncertainty with which the number of cross-links can be determined by chemical analysis, reasonably good agreement is found between the calculated and observed modulus. It is not very often that such a physical property can be predicted theoretically from basic molecular concepts.

The force-length relation expected from Eq. (2) is compared with the actual observed experimental results in Figure 4.6. The theory and experiment are in very good accord for small extension ratios, (L/L_0). Semiquantitative agreement is obtained for extension ratios up to about five. At very large deformations there are indeed some major discrepancies. These discrepancies can be attributed to other factors that are not explicitly taken into account in the simple formulation leading to Eq. (2). Although complete quantitative agreement between theory and experiment has not as yet been attained, a major advance has been made in relating molecular structure to physical properties. The long-range elasticity of rubber-like materials can thus be put on a very quantitative molecular basis. Throughout the analysis the conventional laws and methods of physical science have been used. It was crucial in this development, however, that the underlying structural features of polymer molecules be recognized. This endeavor respresents a classical achievement and demonstrates the great unity of concept that exists in natural science. We must, however, recall with some humility that more than a century elapsed between the observation of the very unique thermoelastic properties of rubber and a molecular understanding of its elastic properties.

Glass Formation

We have found that the existence of long-chain molecules, in disordered conformations, is a necessary condition for the development of long-range elasticity. This is not, however, a sufficient condition. The tacit assumption was made that when an external force is applied, the time necessary for conformational adjustments to take place is very small in comparison with the time scale of the observation, i.e., the time in which the measurements are made. It has, therefore, been implicit in our discussion, heretofore, that the chains are endowed with sufficient internal or segmental mobility to allow the required conformational rearrangements and adjust-

[2] These calculations are not limited to simple extension or stretching. They can be carried out for other types of deformation as well.

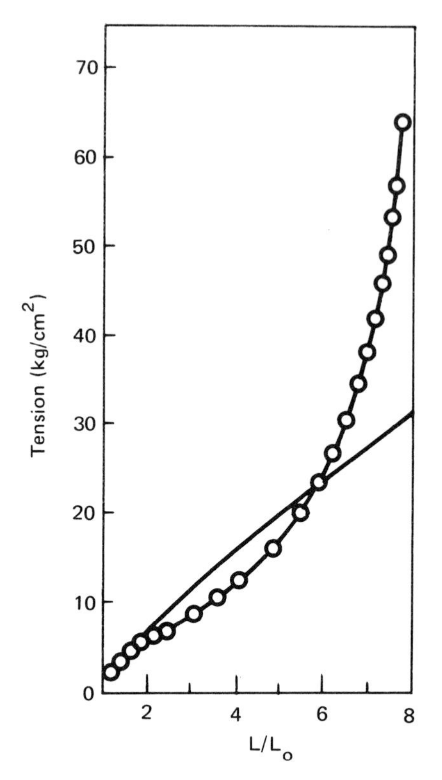

Figure 4.6. Comparison between experiment and theory of the force-length relation of stretched natural rubber. Solid line is the theoretical expectation. Open circle the experimental points. Adapted from L. R. G. Treloar, *The Physics of Rubber Elasticity*, The Clarendon Press, Oxford. Reproduced by permission.

ments to take place. However, it is quite possible that when the temperature is lowered or the pressure raised, the time required for these changes will become comparable with or even greatly exceed the observation time. Under these conditions, despite the fact that the chains are in random conformation and the system is disordered, rubber elastic behavior will no longer be displayed. To be rubber-like, the chains must be able to sample all the available conformations. When this condition is not fulfilled, instead of a flexible, tough, and highly deformable material, one obtains a material that is rigid, very brittle, and easily fractured. The energy imparted by an external force can no longer be dissipated by dimensional changes. Consequently, fracture results. This state is called the glassy state. Although glasses and rubbers clearly represent opposite extremes in mechanical properties, in both cases the chains are in a statistical conformation. Thus, the differences in the properties of the two states do not lie in any fundamental difference in structure or molecular arrangement, but rather in the time scale of the segmental motions and their response to external stress.

Common examples of polymeric glasses are polymethyl methacrylate (Plexiglas) and polystyrene. (Many of our childhood toys were made from polystyrene.) Their fragility and short life span are due to the fact that they are glasses at room temperature. Natural rubber displays all the expected elastic properties at room temperature. However, when cooled to liquid nitrogen temperature, its long-range elastic properties are lost and it becomes very brittle. This change in properties can be dramatically illustrated by taking an ordinary rubber ball and immersing it for a few minutes in a vessel of liquid nitrogen. When the ball is removed from the vessel and thrown against the floor or wall, instead of rebounding in the usual manner it will shatter. At the very low temperature of liquid nitrogen the rubber has become a glass and consequently lost is elastic rubber-like properties. The polymers classified as engineering plastics have relatively high glass temperatures. Therefore, they can retain their form and rigidity at elevated temperatures and thus replace metals in a variety of applications. Glass formation is not, however, restricted to polymeric substances. Monomeric liquids, such as sugars and alcohols, that do not readily crystallize also can form glasses with physical properties similar to those of polymeric glasses.[3] Many sweets—lollipops, for example—are technically glasses and must be handled carefully, lest they fracture.

The transition from a rubber to a glass is reversible—a substance can be transformed from one state to the other by appropriate changes in the temperature or pressure. In addition to the drastic changes that take place in the mechanical properties, other properties are also altered. The fundamental reason is the same, however, in that it takes time for the internal molecular structure to readjust itself to an external change. As the two states are traversed there are discontinuities in the thermal expansion coefficient, in the compressibility, and in the specific heat. In Figure 4.7 we illustrate the relation between the specific volume (reciprocal of the

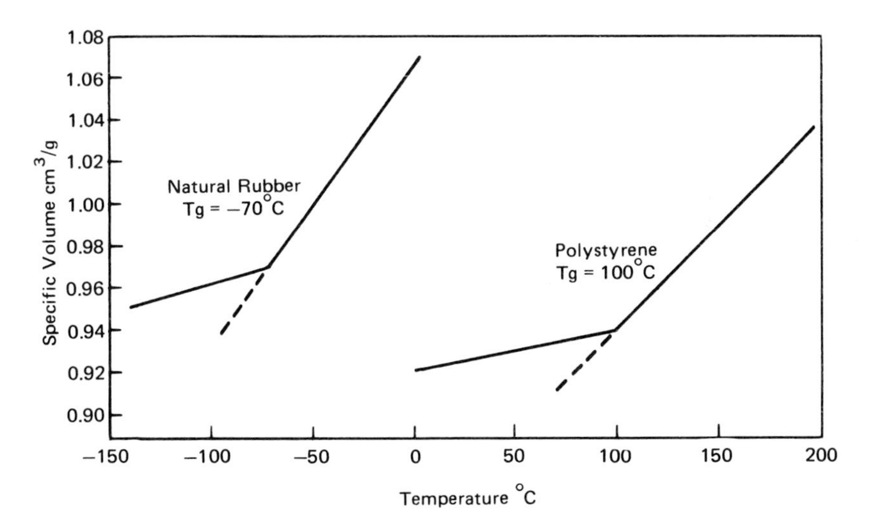

Figure 4.7. Specific volume-temperature relations for polystyrene and natural rubber.

[3] For monomeric liquids it is the time scale of the molecular rearrangements relative to the observation time that causes glass formation.

density) and the temperature as the rubber to glassy state is traversed for two undeformed polymers, natural rubber and polystyrene. Since they are not subject to deformation, both polymers display normal thermal expansion coefficients. However, for each polymer the slope of the volume-temperature curve changes at a well-defined temperature. At this temperature, called the glass temperature, Tg, the specific volumes are continuous but there is a discontinuity or break in the slopes of volume-temperature lines, i.e., in the thermal expansion coefficients. The thermal expansion coefficient in the rubbery state is usually two to three times greater than in the glassy state because of the greater molecular mobility. If we could carry out each observation below Tg for an exceedingly long time then we would not expect to see this discontinuity. Instead, the dashed lines would be observed. However, over the time intervals that man can be expected to make observations, specific volume measurements serve as both a very reliable and convenient method of locating the glass temperature.

All polymers experience very similar changes in properties when they undergo the glass to rubber transition. The only difference is the temperature at which these changes occur. As a further example, we examine in Figure 4.8 the modulus of elasticity for natural rubber and polystyrene as a function of temperature. The modulus is represented on a logarithmic scale in order to be able to plot on one graph the large range in values. The transition region, which encompasses a 10- to 20-degree range, is characterized by about a 1000-fold change in the modulus. The modulus is of course a reflection of the extensibility of the material. We should note that the transition temperature for these two polymers, as determined by the large change in a physical property, is the same as is found from the specific volume measurements shown in Figure 4.7. The shapes of the two curves in Figure 4.8 are

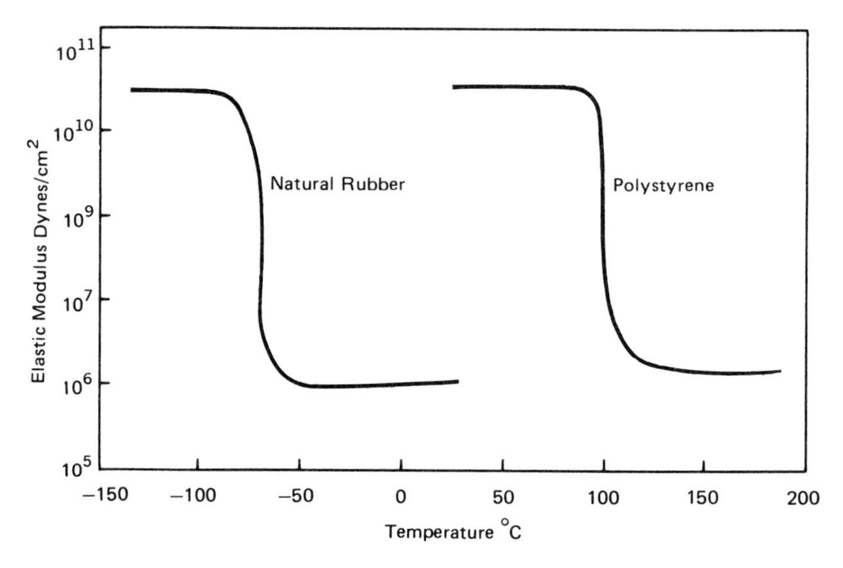

Figure 4.8. Elastic modulus of polystyrene and natural rubber as a function of temperature.

virtually the same when cognizance is taken of the difference in glass temperatures. Below the glass temperature the modulus of elasticity is 10^{10} dynes per cm^2 for both polymers. It remains essentially constant with increasing temperature until the transition temperature. At this temperature there is a precipitous decrease in the modulus as the sample becomes a rubber. The modulus then remains essentially constant at about 10^7 dynes per cm^2 in the rubbery state. Figure 4.8 illustrates quite vividly the consequences of the dual personality of a polymer chain even though it is in a statistical conformation.

Chemical Structure and Glass Formation

Table 4.1 sets forth the values of Tg for some representative polymers. There is a wide variation in the values of the glass temperatures, from -120°C for polydimethyl siloxane to more than 200°C for the polysulfone. The nature of the chemical repeating unit plays a very important role in determining the temperature at which the transition between the rubbery to glassy state occurs. The major chemical factors appear to be the flexibility of the chains (i.e., the relative freedom of rotation about the single bonds), the bulkiness of side groups, and the intermolecular interactions between substituent groups. The more flexible the chain, the lower the glass temperature. Intense polar interactions and bulky side groups tend to raise this temperature. The first five polymers listed in Table 4.1 are usually rubbers or elastomers at room temperature. Their rubber-like properties clearly result from the fact that their glass temperatures are located well below their use temperatures. Otherwise, the necessary response of the chain conformation to external stress would not occur and the long-range elastic properties would be lost. On this basis, poly (di-

Table 4.1. Glass Temperatures, Tg, of Polymers

Polymer	Tg°C
Polydimethylsiloxane–silicone rubber	-120
Polybutadiene	-85
Polyisobutylene–butyl rubber	-70
Poly *cis*–1,4–isoprene–natural rubber	-70
Polychloroprene–Neoprene	-50
Polydichlorophosphazene	-63
Polyvinyl chloride	+80
Polymethyl methacrylate–Plexiglas	+100
Polystyrene	+100
Poly-α-methylstyrene	+130
Polyethylene terephthalate–Dacron, Mylar	+60
Polyhexamethylene adipamide–6,6–Nylon	+54
Polypropylene	-20
Polydimethyl phenylene oxide	+207
Polybisphenol-A-carbonate–Lexan	+150
Polyether sulfone	+230

chlorophosphazene) and similar polymers can serve as completely inorganic rubbers. For the same reason polymers with high glass temperatures are not usually suitable as elastomers. However, if used above their respective Tg's they would display the usual rubber-like behavior. Such common plastic materials as polymethyl methacrylate, polystyrene, and polyvinyl chloride have glass temperatures well above their use temperature. The last three polymers listed in the table have exceptionally high glass temperatures. As previously mentioned, this is the principal reason for their use as engineering plastics.

Plasticizers

The glass temperature of a polymeric system must naturally be altered by the incorporation of structural or molecular features that influence chain mobility. Copolymerization effects the glass temperature, as does the addition of nonvolatile monomeric liquids. In the simplest case, the incorporation of a co-unit into the chain results in a monotonic change in the glass temperature between the values for the two corresponding homopolymers. Thus, major alterations in properties can be attained, relative to the use temperature, by this kind of compositional change. The introduction of a pure liquid into a polymeric system lowers the glass temperature. Such liquids are called plasticizers. They can facilitate segmental motions and therefore act as internal lubricants. An important practical application of this principle can be illustrated in the uses of polyvinyl chloride. According to Table 4.1, the glass temperature for the pure, undiluted polymer is 80°C. Consequently, at room temperature it is a brittle, inelastic material, which finds little use in its pure form because of these undesirable physical properties. Its flexibility and elastic properties can be enhanced, however, by the addition of a plasticizer that lowers the glass temperature below room temperature. Certain desirable physical properties are then given to the polymer. It has become, for example, the major constituent of plastic raincoats, garden hoses, and long-playing phonograph records. This polymer has found extensive application in these examples because the use temperature is located just above the glass temperature. The system is neither completely rigid on the one hand nor does it exhibit rubber-like behavior on the other. There are, however, certain disadvantages in this use. Although nonvolatile plasticizers are used, the liquid molecules will slowly migrate or diffuse to the surface. As a result, the enhanced segmental mobility will be lost and the article will gradually stiffen and become brittle. After a while the item will no longer be useful for the purpose designed. We have all experienced the stiffening and cracking of plastic raincoats as well as the failure of garden hoses. The use of improper liquid cleaner for phonograph records leaches out the plasticizer and causes embrittlement and eventual fracture.

In this chapter we have discussed the major properties of polymers when in the disordered state. The property of long-range elasticity, unique to long-chain molecules irrespective of their chemical constitution, can be explained in a straightforward manner when the conformational characteristics of the chains in this state are recognized. However, we have also seen that there is also a characteristic temperature at which the rubber-like properties are lost and the same polymer becomes a brittle glass. The chemical nature of the chain repeating unit, which determines the glass temperature, ultimately determines the end-use of the polymeric material. As we have noted in Chapter 3, polymer chains can also exist in highly ordered conformations. The properties characteristic of this state are quite different from those we have just discussed and form the subject of the next chapter.

Chapter 5

CRYSTALLINE POLYMERS AND FIBERS

Crystal Structures

In this chapter we shall examine the properties of a collection of polymer molecules in the crystalline state. In this state each chain adopts the same ordered conformation. Because of the high degree of molecular order the structure in this state is very similar to that of monomeric solids. It is virtually axiomatic, and has been amply demonstrated, that polymers with a sufficiently regular chain structure will crystallize. The type of chain irregularities that retard or even suppress the crystallization process are chemically different units such as copolymer units, branch points, and chemically identical units that are geometric or stereo-isomers. It should not surprise us that the drastic difference in the chain conformation is reflected in some major differences in properties as compared with the disordered state. For example, the modulus of elasticity increases anywhere from 1,000- to 10,000-fold, reflecting the loss of long-range elasticity in the ordered state. Thus in the crystalline state a polymer chain is not easily deformed.

In Chapter 3 we found that when an isolated long-chain molecule adopts an ordered conformation the geometric form that evolves is rod-like and asymmetric. The crystalline state, which is a collection of molecules in such conformations, represents, in addition, order in three dimensions. Thus, each chain is not only conformationally ordered but the molecular axes are also parallel to one another. The axes are so arranged that the chain substituents form repetitive, regular planes in directions normal to the chain axis. A structural description of the crystalline state requires not only a very definite chain conformation but also a well-defined mode of chain packing. The necessary information about the molecular arrangements in polymer crystals is obtained from X-ray diffraction measurements. The experimental techniques and theoretical analysis are the same as are used in the structural determination of crystals of monomeric substances. The molecular arrangement can be described by a unit cell. This is the structural unit that is repeated in all directions and thus produces a three-dimensional lattice. Some typical ordered chain structures, all of which represent conformations of minimum energy, have been described in Chapter 3. We shall show some three-dimensional structures in order to emphasize some of the packing characteristics. Figure 5.1 illustrates the crystal structure of linear polyethylene. The planar zigzag chain conformation is readily apparent. The repeating distance along the chain axis, which is 2.53 Å, represents the length of one zigzag (two carbon atoms). The unit cell is a rhombus, as is indicated in the figure. The very specific regular disposition of the chains can also be seen. The cross-sectional area of the unit cell is 36 Å^2. The two chains per unit cell are thus positioned fairly far apart from one another.

A more complex ordered array is found for crystalline natural rubber illustrated in Figure 5.2. We should note parenthetically that this molecule, *cis* polyisoprene, which represents the classical rubber or elastomer when in the disordered state, will

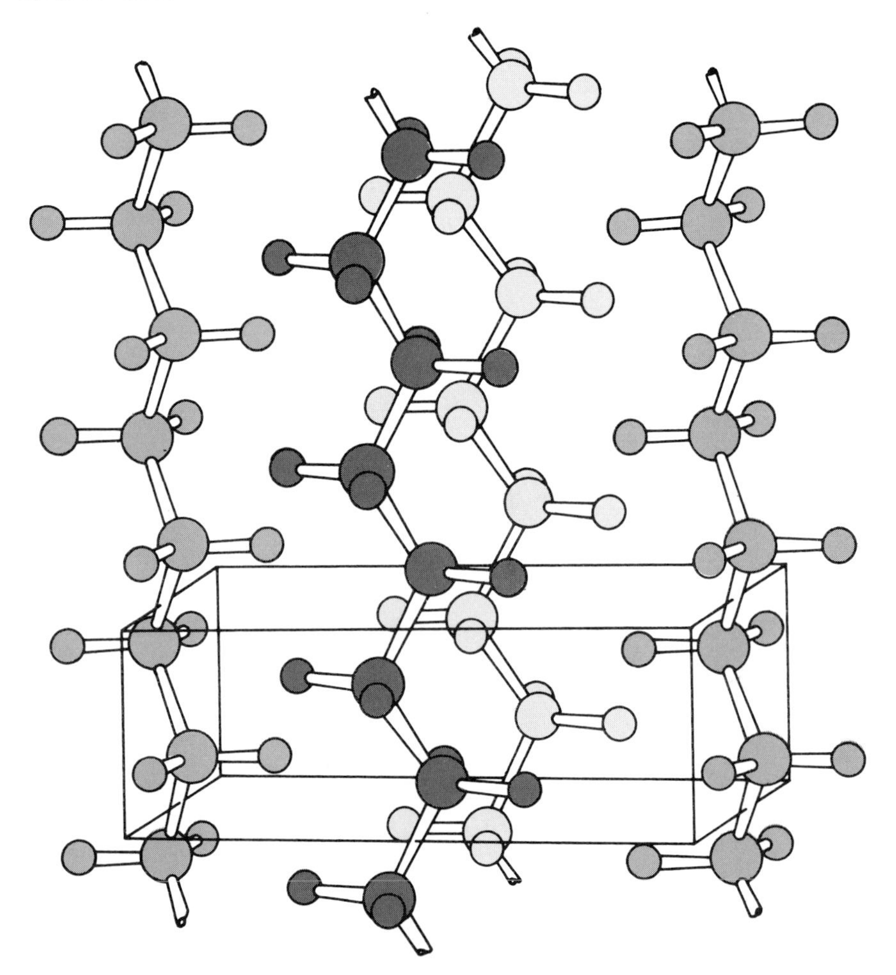

Figure 5.1. The crystal structure of polyethylene. From C. W. Bunn, *Chemical Crystallography*, Oxford University Press. Reproduced with permission.

crystallize or order when held for sufficiently long times well below room temperature, but above its glass temperature. The regularity of the chain structure is again evident. The *cis* configuration is maintained repetitively about the double bonds. The repeat distance along the chain axis is 8.1 Å, which indicates that there are two isoprene units per geometric repeat. Here the chain is not in an extended conformation. The unit cell is monoclinic in this case so that the planes making up the unit cell do not meet at right angles.

The ordered chain arrangements for two aliphatic polyamides have been given in Figure 3.15. Here, the interchain distances are governed to a large extent by the strong propensity for hydrogen bonding between the O atom from the C=O group of one chain and the H atom from the N–H group of an adjacent chain. The ordered structure for another fiber forming polymer, the polyester polyethylene terephthalate, is shown in Figure 5.3. The chain repeat distance is 10.75 Å, which is only

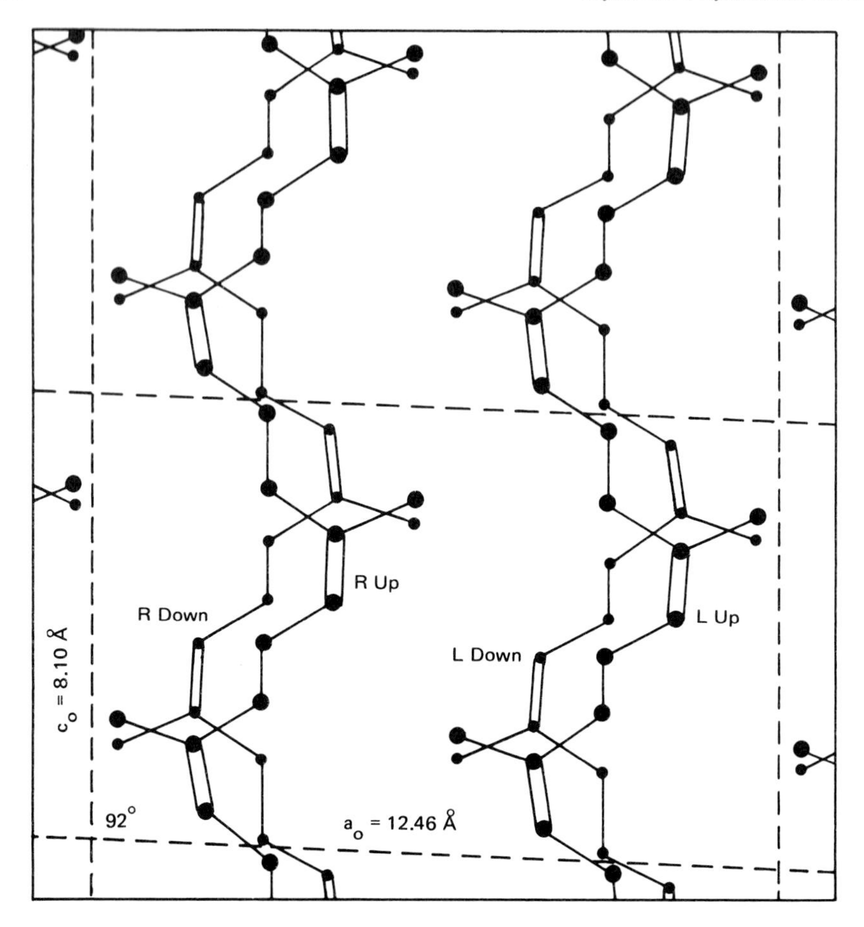

Figure 5.2. The chain arrangement in natural rubber crystals. From C. W. Bunn, Proceedings of the Royal Society *180A*, 40 (1942). Reproduced with permission.

slightly less than the requirements for a fully extended chain. The chains are, therefore, in a nearly planar conformation. In the unit cell the edges are of unequal length and the angles are different. There is one chain in the monoclinic unit cell.

The structures illustrated above were selected from among many examples primarily to demonstrate the principle that long-chain molecules can be organized into a well defined three-dimensional ordered array.

Crystallization Process

At this point we may very easily be puzzled as to how such ordered structures can be formed from a polymer melt. There are two quite different modes of crystallization (or three-dimensional ordering) that depend of the initial chain conformation. Prior to crystallization, chains can exist in their disordered conformation, which

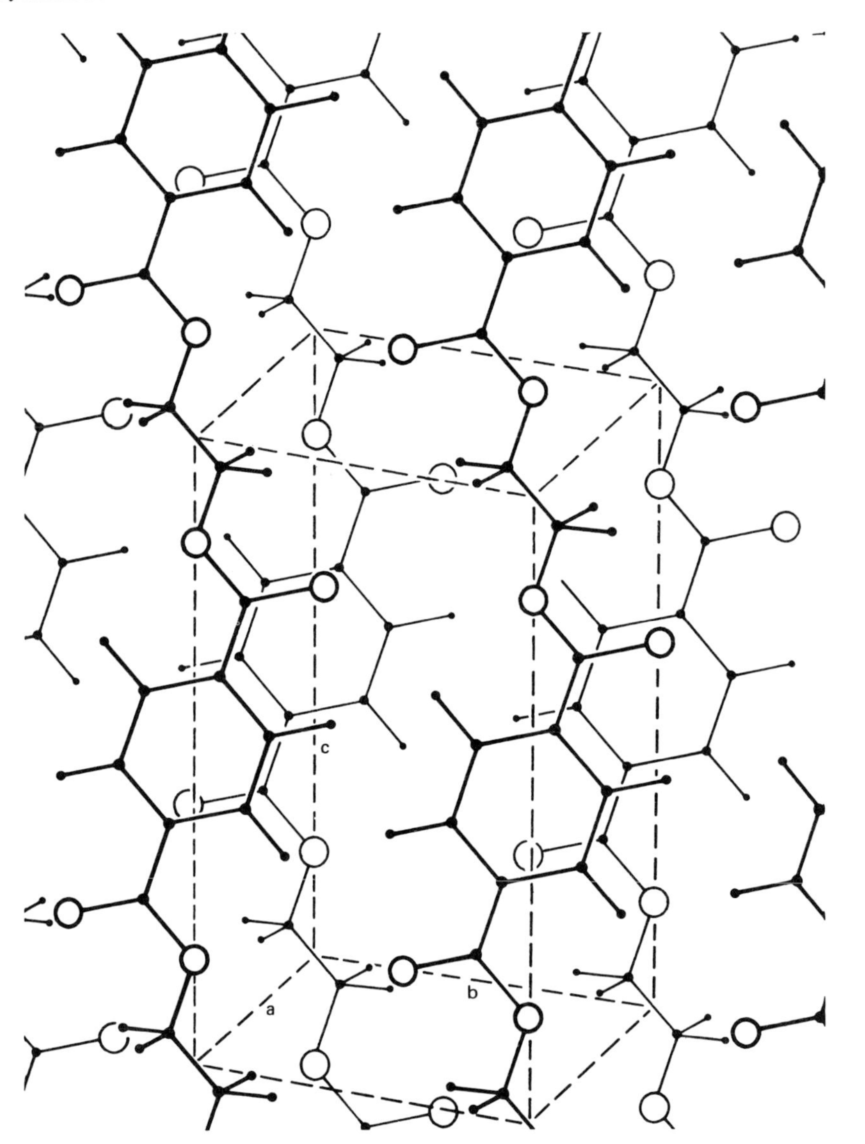

Figure 5.3. The chain arrangement in polyethylene terephthalate crystals. From R. P. Daubeney, C. W. Bunn and C. J. Brown, Proceedings of the Royal Society *226A*, 531 (1954). Reproduced with permission.

corresponds to the liquid or molten state. Alternatively, as we have previously discussed, certain types of individual chains can have highly ordered conformations and thus form a liquid crystal or nematic state. We consider first the situation where the chains are initially in random conformation. In the molten state the chains are in disordered conformations with their characteristic irregular structures and a great deal of intermolecular entanglement. Despite the differences that must exist in the mechanistic nature of the crystallization of polymers as compared to monomers, the process formally the same for both kinds of substances. The new state or phase must be initiated within the body of the parent liquid. This initiation process is called nucleation. The nuclei, or crystal embryo as they are sometimes called, subsequently grow into larger mature crystals. Both of these processes have been identified in polymer crystallization. The crystallization from most monomeric liquids takes place at very rapid rates at temperatures just slightly below the melting temperatures. The crystallization conditions are thus very close to that for equilibrium between the crystal and liquid. For example, the *n* paraffin, $C_{94}H_{190}$, tetranonacontane, crystallizes very rapidly at temperatures just infinitesimally below its melting point. On the other hand, for the polymeric analogue of this compound, linear polyethylene, as well as for all other polymers, crystallization must invariably be conducted at temperatures well below the melting temperature in order that the process can proceed at an appreciable rate. Depending on the specific polymer, crystallization temperatures anywhere from 15° to 50° below the melting temperature are necessary. Consequently, polymer crystallizations take place under conditions well removed from equilibrium. This requirement introduces complications into the understanding of the crystalline state of polymers, in the interpretation of the corresponding properties, and thus sets it apart from crystals of monomeric substances.

The concept of polymer crystallization outlined so far is an obvious oversimplification. We have implied that all of the long-chain molecules, each of which is comprised of many thousands of chain-repeating units, achieve a regular conformation over their complete molecular length. Furthermore, the molecules must be aligned in a parallel array with one another. Although this situation represents an idealized equilibrium requirement, the kinetic and mechanistic difficulties that must be overcome to achieve this condition are enormous. The difficulty that a chain would experience even to extricate itself from the highly disorganized molecularly entangled melt is easily understood. This process must take a great deal of time. Even if the separation of the chains were accomplished, the attendant molecular ordering would be a kinetic, time-dependent process. In addition, the fact that polymer crystallization is invariably conducted at temperatures well below the melting temperature will contribute further to the nonequilibrium character of the final state. Because of these problems a compromise structure must be reached in any actual crystallization. A complete transformation from the disordered to crystalline state is rarely, if ever, achieved. Only a portion of a chain will adopt the required ordered conformation. Depending on the polymer type, molecular weight, and crystallization conditions, the percentage of the material crystalline will vary from 30 to 90 percent. The failure of the level of crystallinity to reach very high values at all times, despite the great thermodynamic driving force urging the chains to do so, can be attributed to the intractability of a long chain, particularly when a portion of it becomes embedded in a crystallite. Entangled portions of the molecules that are trapped in the regions between crystallites will not be able easily to undergo the conformational rearrangements necessary for the deposition of an ordered structure on the surface of a crystallite.

A real polymer system is therefore polycrystalline, i.e., made up of many small crystallites. The crystallites have dimensions (in the chain direction) that are on the order of several hundred angstroms, and they are interspersed within a matrix of disordered or amorphous chain units. We should, therefore, not be surprised that the properties depend not only on the fact that the system is crystalline, but also on the amount of crystallinity and on the crystalline morphology. By morphology we mean the size and shape of the crystallites, the chain arrangement within crystallites, the relative arrangement of the crystallites with one another, called the supermolecular structure, and the detailed nature of the amorphous zone.

Crystallization can be carried out with the application of an external force, so that the transformation takes place from a deformed or rubber elastic melt. In contrast, the crystallization could also be conducted without the application of an external force. In this case the transformation would take place from an undeformed melt. Although here each of the crystalline regions represents a highly ordered domain, the regions themselves are randomly arranged relative to one another. The fact that such structures actually exist can be demonstrated by the methods of X-ray diffraction. Some typical X-ray patterns are given in Figure 5.4. In Figure 5.4(a) the X-ray pattern is only a very diffuse halo. This kind of pattern is characteristic of a noncrystalline substance. The particular example is for natural rubber at room temperature, which is in the disordered or amorphous state. However, when a polymer is crystallized merely by cooling, with no deformation being imposed, a series of intense, sharp, concentric rings are found. An example of an X-ray pattern for this type of crystallization is given for linear polyethylene in Figure 5.4(b). This pattern is known as a powder diagram. It has the same characteristics as one obtained from a very fine crystalline powder of a monomeric substance. In this case the small crystallites are also randomly arranged relative to one another. From the position of the rings on the film the structure of the unit cell can be derived by conventional methods.

When a polymer is crystallized under a constant external force, as for example in simple extension, the morphological picture is changed. Because the crystallization is conducted from a deformed melt, the crystallites are now preferentially oriented in the stretching direction. The chain axes are thus directed along some macroscopic dimension in space. The characteristic X-ray diffraction pattern is also altered, as is shown in Figure 5.4(c) for some typical oriented crystalline polymers. These polymers include natural rubber, linear polyethylene, and the fibrous protein collagen. Instead of concentric rings, we now observe a set of discrete spots arranged in a very regular pattern. The fact that one observes spots instead of rings is indicative of crystallite orientation because the diffracting planes of the crystals must be biased relative to the X-ray beam. A quantitative analysis of such patterns yields the details of the orientation as well as the unit cell structure.

Similar oriented systems can also be obtained by other processes. In one case the initial state is that of the liquid crystal. We shall discuss crystallization under these conditions subsequently. In another case of importance, one starts in the undeformed crystalline state. An external force of sufficient intensity is applied to deform the disordered regions. As a consequence, the connected crystalline regions are oriented in the stretching direction. In many practical cases a high state of crystallite orientation can be achieved more easily by this process, which is technologically called drawing.

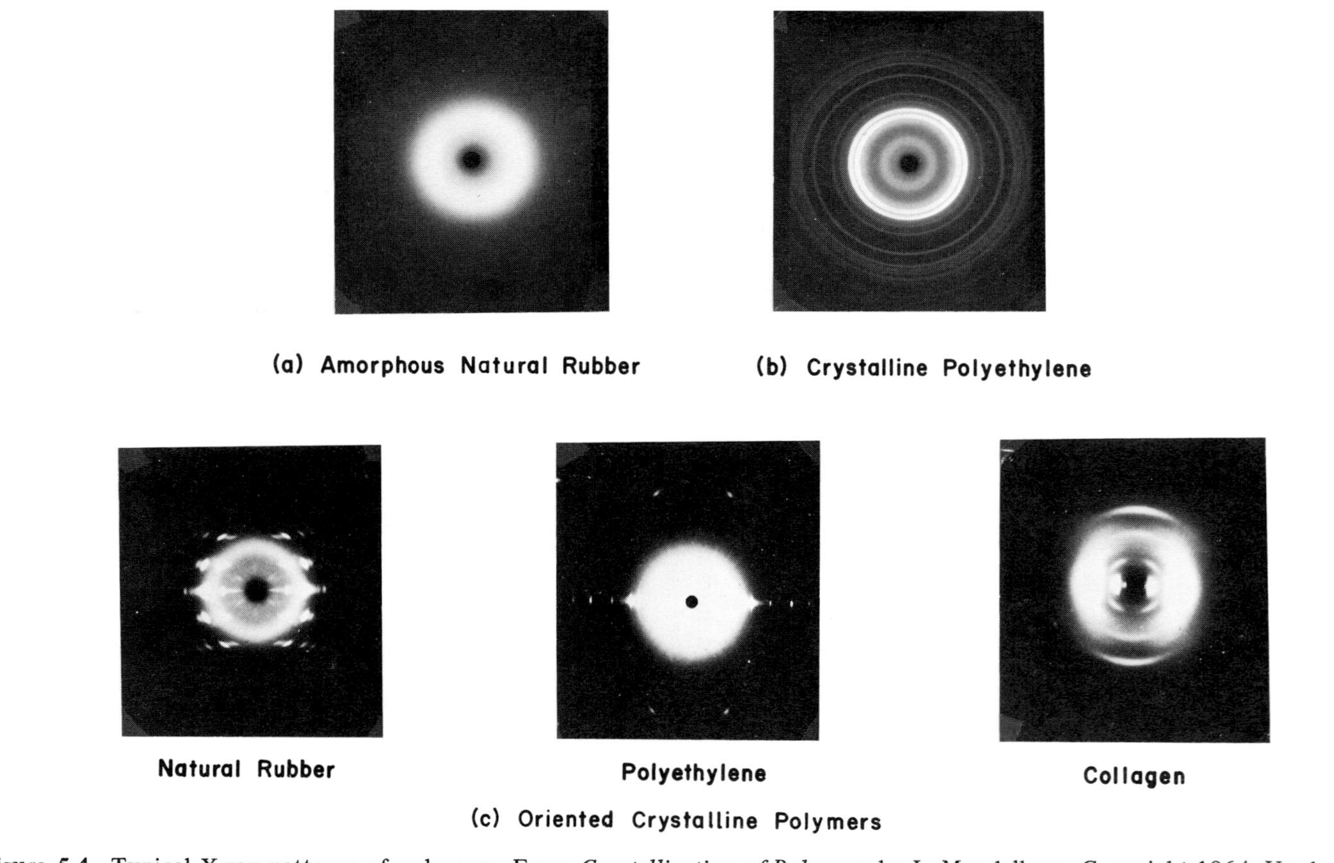

Figure 5.4. Typical X-ray patterns of polymers. From *Crystallization of Polymers* by L. Mandelkern. Copyright 1964. Used with permission of McGraw-Hill Book Company.

Undeformed Crystallization–Morphology and Properties

Crystallization in the undeformed state, i.e., from chains in random conformation, can be carried out over the complete composition range from very dilute solutions to the pure bulk polymer. Crystals grown from dilute solutions–i.e., polymer concentrations of the order of one percent or less–display a very characteristic morphological form. Under the electron microscope the crystals are found to be in the form of thin plates or lamellae. A typical electron micrograph of such crystals is given in Figure 5.5. This morphological form is found for all homopolymers crystallized from dilute solution. The lamellae are about 100 Å thick and extend to several microns in the transverse directions. Given these dimensional requirements we would intuitively expect that the chains within the crystallite would be arranged so that their molecular axes lie parallel to the wide faces of the lamellae. However, standard methods of studying orientation have established rather surprisingly, but convincingly, that the chain axes are directed perpendicular to the wide faces. Since the platelet thickness is only of the order of 100 Å, and since such crystals are formed from chains of very high molecular weight, the crystallite thickness must represent only a very small fraction of the extended length of a typical molecule. Conse-

Figure 5.5. Election micrograph of polyethylene crystals formed in dilute solution. From *Crystallization of Polymers* by L. Mandelkern. Copyright 1964. Used with permission of McGraw-Hill Book Company.

quently, a chain must traverse a given crystallite many times in order to simultaneously satisfy both the dimensional and orientation requirements. Hence the platelets must consist of some type of folded polymer chains. The details of the molecular structure of a polymer chain located within such a platelet has been a matter of widespread discussion for many years.

The very regular, smooth appearance of the crystals under the electron microscope led initially to the concept that the chains were regularly folded or pleated. A regularly structured, smooth interface with hairpin-like bends was assumed. Although this feat is sterically possible for most chains, detailed physical-chemical studies of such crystals have shown that despite outward appearances the surfaces are not smooth or regularly structured. Instead, a disordered overlayer exists, as is schematically depicted in Figure 5.6. The required chain orientation and lamellae structure is still maintained, but the crystalline sequences are randomly connected by loops of random length. About 10 to 15% of the chain units form the disordered interfacial layer so that the system is not completely crystalline. This disordered layer, well established by physical measurements, cannot be resolved by the electron microscope, but is consistent with the direct observations. A more quantitative description of the details of the lamellar interfacial structure is being developed.

The characteristic lamella morphology of crystals formed in dilute solution, with its unexpected chain orientation, is of interest not only in itself but also has very great significance with respect to the structure of polymers crystallized from the

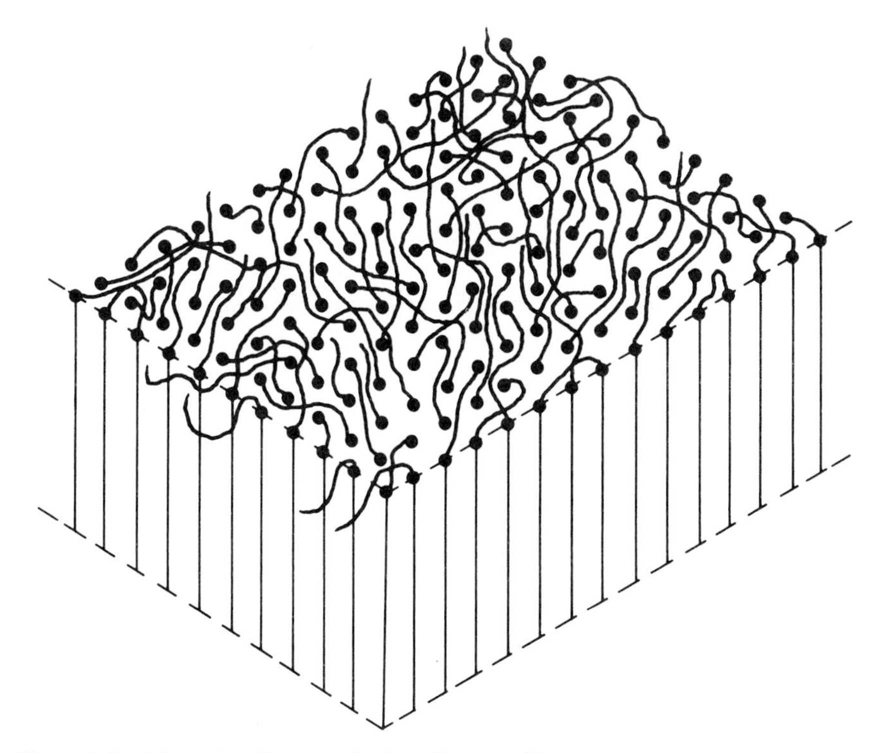

Figure 5.6. Schematic diagram of a lamella crystallite.

Figure 5.7. Electron micrograph of a replica of a fracture surface of linear polyethylene. Scale 1 mm = 370 Å units. From Journal of Polymer Science A-2 *4*, 385 (1966). Reproduced with permission.

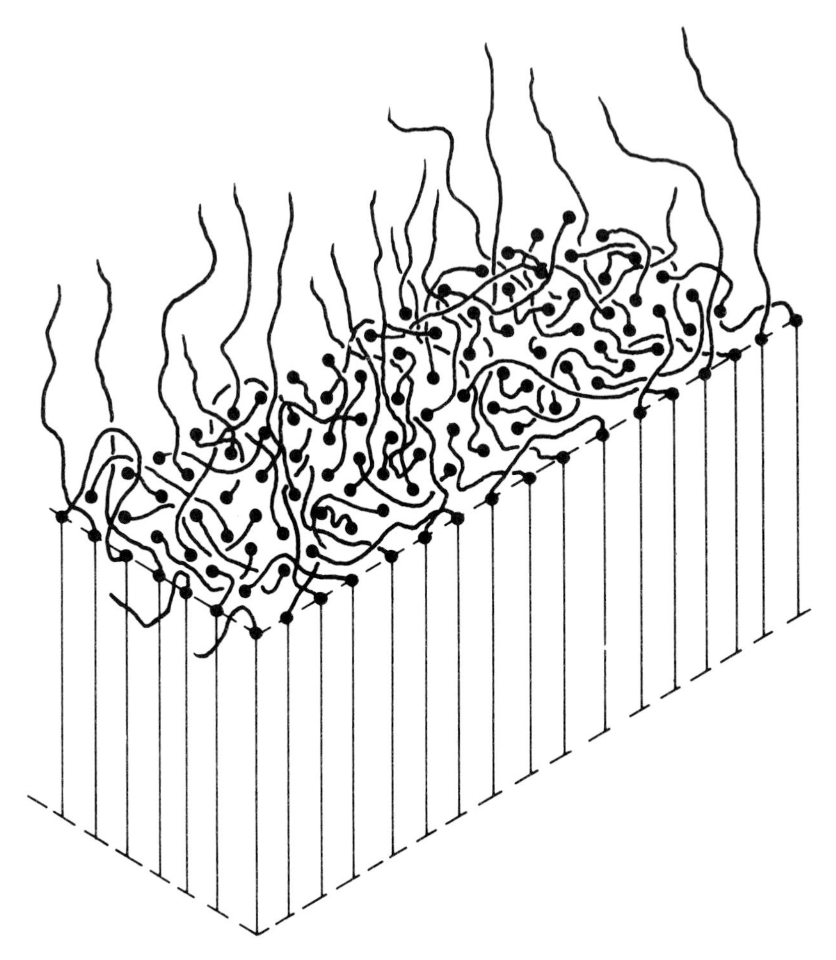

Figure 5.8. Schematic representation of morphology of bulk crystallized polymer.

pure melt. A variety of experimental techniques, including electron microscope observations, have established that a lamella crystallite is also the primary morphological form for bulk crystallized systems. A typical electron micrograph, showing a stack of thin lamellae with large lateral dimension, is given in Figure 5.7. For bulk crystallized polymers the lamella thickness can vary from several hundred to several thousand angstroms. Based upon a large number of detailed studies of many properties, the primitive crystallite structure of a bulk crystallized polymer can be represented as in Figure 5.8. The vertical lines represent the chain sequences within the crystalline regions, which are in an ordered conformation. The ordered structures are either planar zigzag, or one of the appropriate helical forms. As they emerge from the basal plane of the crystal some of the chains will return to the crystallite of origin but not necessarily in juxtaposition. Others will leave the vicinity of the crystallite and will form a disordered amorphous or interzonal region which eventually joins a neighboring crystallite. The chain units in the interzonal region, which connect one crystallite with another, are in a disordered conformation. An ill-defined, disordered, high energy, and irregularly structured interfacial zone is also formed. The relative proportion of each region depends on the molecular weight and the crystallization conditions, particularly the crystallization temperature. Thus, in the crystalline state, perhaps realistically called the partially crystalline state, chain sequences that are organized into a highly ordered array and chain units in disordered conformation coexist and are connected with one another. We can therefore expect a wide range in properties as the proportions of these regions are varied for a polymer of the same chemical constitution.

Before we examine the properties of crystalline polymers in somewhat more detail, it is of interest to describe the main types of supermolecular structures that are observed. Here we are concerned with the arrangement and organization of the crystallites into larger defined structures. We must recognize, however, that supermolecular structures are not always found in crystalline polymers. Very high molecular weight homopolymers, and copolymers under certain crystallization conditions, still form well-defined lamella-like crystallites. In these cases, however, the crystallites are randomly arranged relative to one another so that no organized superstructure develops.

The superstructures that are formed can be directly observed by electron and light microscopy and indirectly inferred by other methods. A collection of lamellae organized into rod or sheet-like structures can be achieved by crystallizing low and modest molecular weight samples at high temperatures. An example of this type of superstructure is shown by the transmission electron micrograph of Figure 5.9. In this micrograph the light areas are the lamellar crystallites while the intervening dark regions are the stain absorbing, disordered amorphous regions.

By far the most common type of organized structure that is found in crystalline polymers is a spherulite. These structures are usually developed by rapid crystallization. An electron micrograph illustrating a typical spherulite that is found in linear polyethylene is shown in Figure 5.10. In a very rudimentary sense a spherulite can be considered to be a spherical aggregate of crystallites. The electron microscope shows that the lamellae are organized into fibrillar structures that radiate from a common center. Spherulites can very often be grown to very large sizes. Diameters in the millimeter range are not uncommon in some polymers. Spherulites are also birefringent and are thus easily observed under the polarizing microscope. A typical light micrograph, under polarized light, for a spherulite in poly(ethylene oxide) is given in Figure 5.11. The magnification in this micrograph is only 125 times, so that the spherulites are indeed quite large.

From our brief excursion into examining the supermolecular structural aspects of morphology, we have seen that a polymer with the same chemical repeating unit can yield different structures depending on molecular weight and crystallization conditions. In fact, major changes can be achieved with exactly the same sample by just varying the crystallization conditions.

We are now in a position to examine in more detail the properties of crystalline polymers. Figure 5.12 illustrates how some of the properties of linear polyethylene change with the amount of crystalline material. For this polymer the fraction of material crystalline can be varied from about 0.5 to 0.9 by control of the molecular

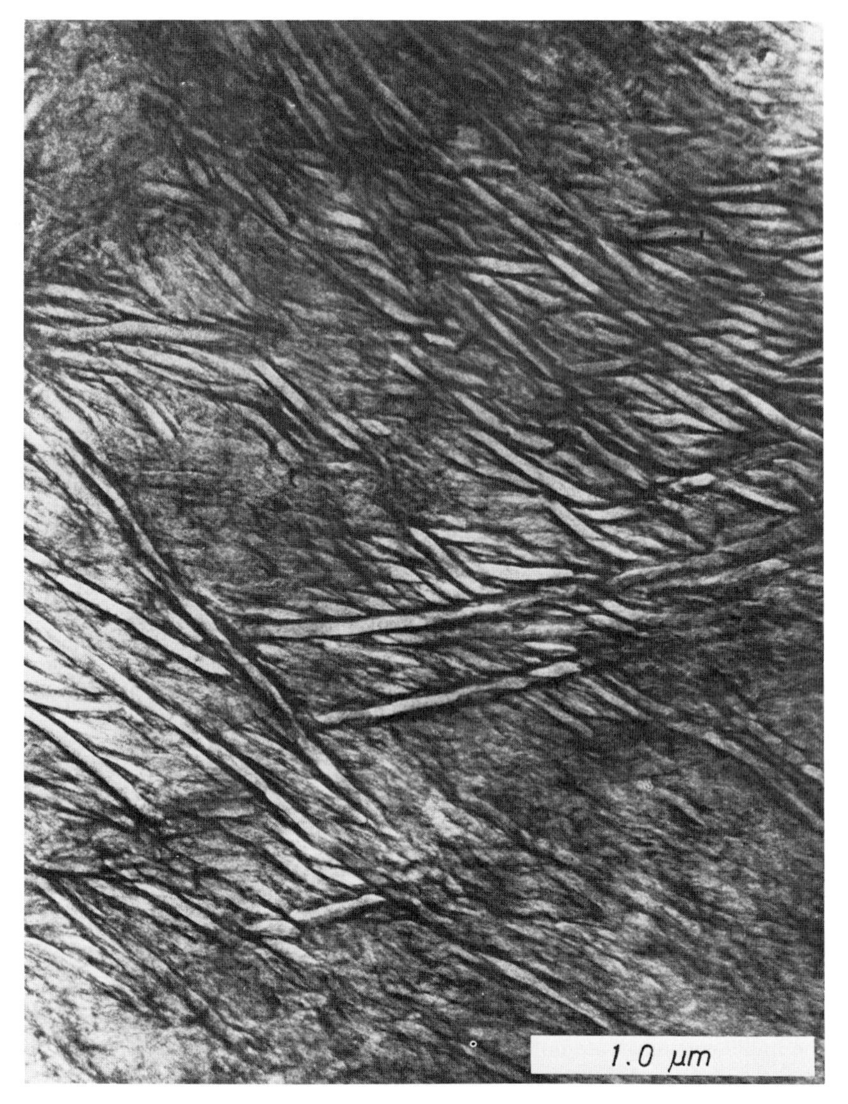

Figure 5.9. Transmission electron micrograph showing sheet like structure in crystalline linear polyethylene. From I. G. Voigt-Martin, E. W. Fischer and L. Mandelkern, *J. Poly. Sci., Poly. Phys. Ed. 18 2347 (1980)*. Reproduced with permission.

weight and crystallization temperature. The density at room temperature ranges from about 0.99 $g \cdot cm^{-3}$ to 0.92 $g \cdot cm^{-3}$. The former value is quite close to the density of the unit cell and thus represents an almost completely crystalline polymer. The latter value, however, represents about only half the material being crystalline. Similar changes are observed in another thermodynamic property, the heat of fusion. Here the values range from about 68 cal/gram to about 25 cal/gram. This

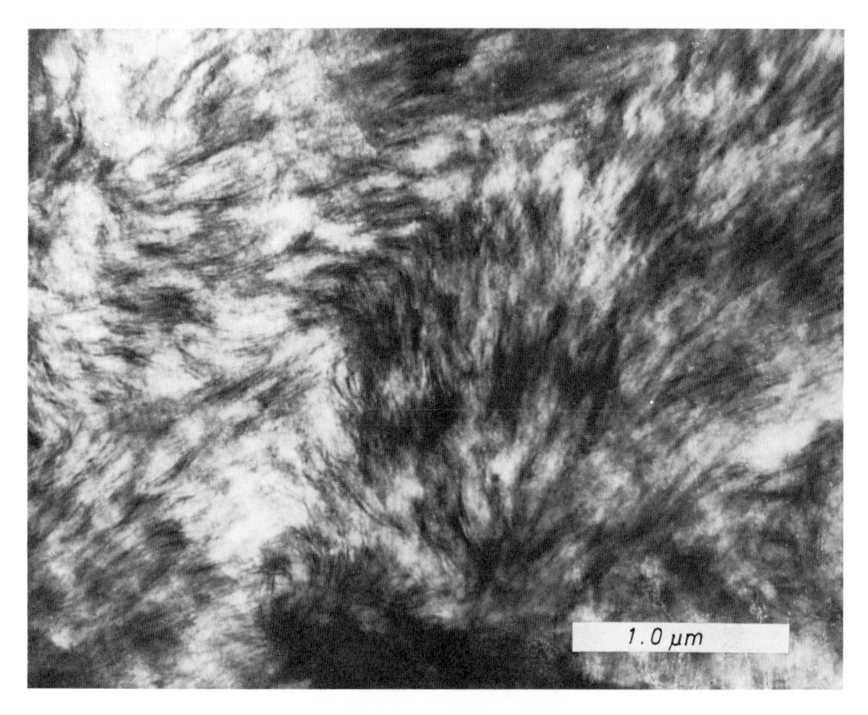

Figure 5.10. Transmission electron micrograph showing spherulitic structure in crystalline linear polyethylene. From I. G. Voigt-Martin, E. W. Fischer and L. Mandelkern, *J. Poly. Sci., Poly. Phys. Ed. 18 2347 (1980).* Reproduced with permission.

range in values for the density and heat of fusion represent very large changes for an identically constituted chemical system. Such large variations in properties are unheard of for low molecular weight compounds. In crystalline polymers this effect is obviously due to the two kinds of chain structures that are present and to the drastically different contribution each makes to a given property.

Besides the thermodynamic properties, changes are observed in other physical properties with the crystalline content. For example, samples with a very high level of crystallinity are very brittle and are easily fractured at room temperature. They thus can find very little practical use. As the level of crystallinity is reduced, the polymer becomes tougher and more flexible. It is no longer brittle and is not easily fractured except when cooled below its glass temperature. When the fraction of crystalline material in linear polyethylene is reduced below about 0.85 it becomes a material with a wide range of applications. The change in the elastic modulus with the level of crystallinity is also shown in Figure 5.12. This quantity measures the flexibility of a material. There is about a five-fold change in the modulus as the fraction crystalline varies from 0.85 to 0.60. The mechanical, physical, and elastic properties of crystalline polymers are also modified by the different types of supermolecular structure.

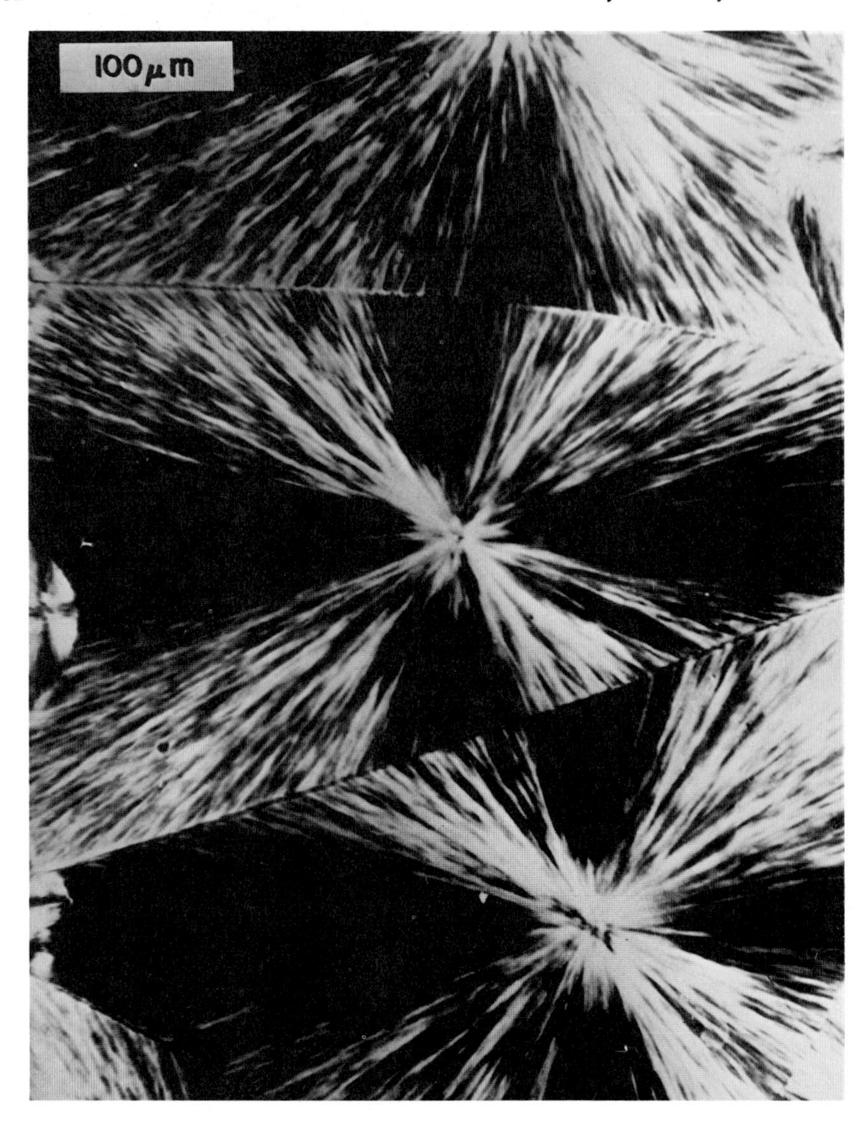

Figure 5.11. Polarized light micrograph of spherulites in poly (ethylene oxide). Magnification 125x. Courtesy R. C. Allen.

Melting Process–Underformed Crystallization

The transformation from the crystalline to the amorphous or disordered state follows a characteristic pattern for all polymers. This behavior is very similar, if not identical to the melting of monomeric solids. An example of the fusion of a high molecular weight fraction, which was crystallized very carefully and then heated very slowly, is given in Figure 5.13. The specific volume, which is directly related to

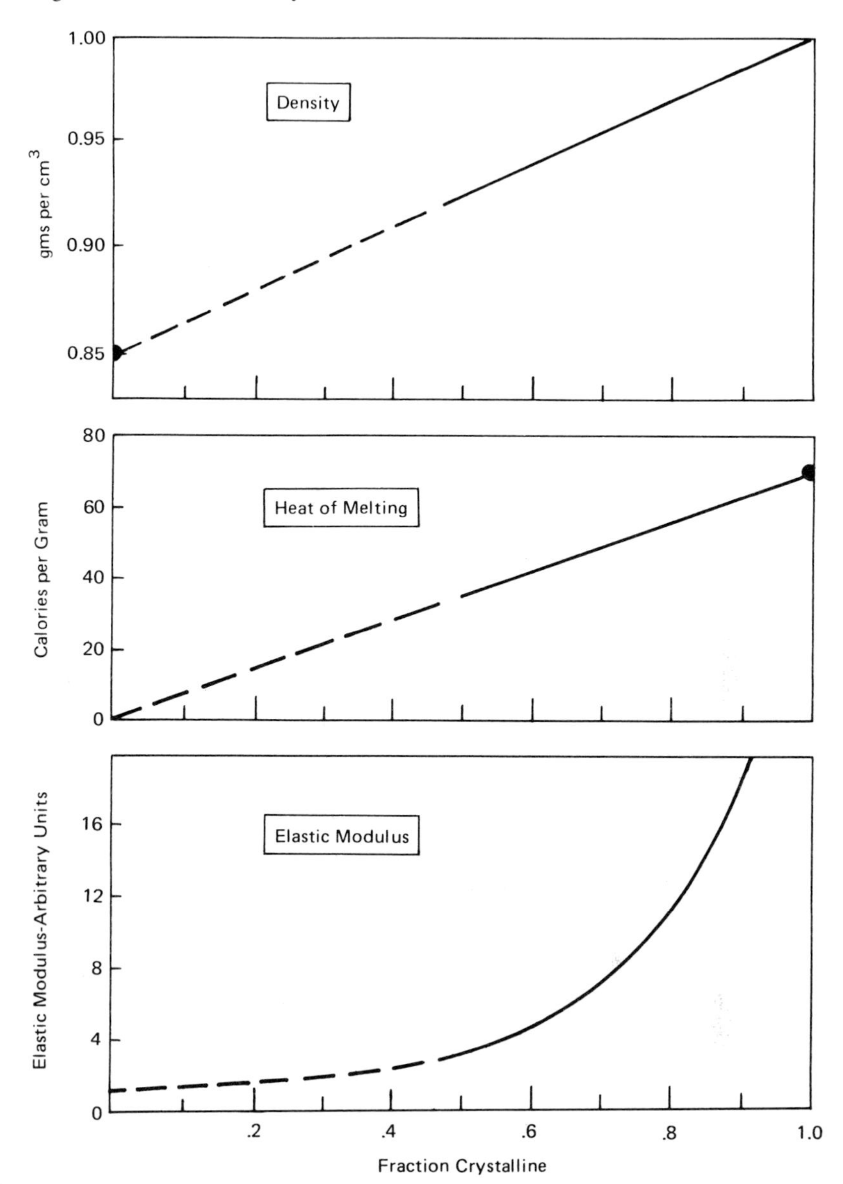

Figure 5.12. The change in the properties of polyethylene as the fraction of material crystalline is varied.

the fraction crystalline, is plotted as a function of temperature. At the lower temperatures, a linear volume-temperature relation is found that is typical of the crystalline state; at higher temperatures, the volume-temperature coefficient, although still linear, has a much higher value, which is characteristic of the liquid state. About 80% of the transformation takes place over about a 2° temperature interval as is shown

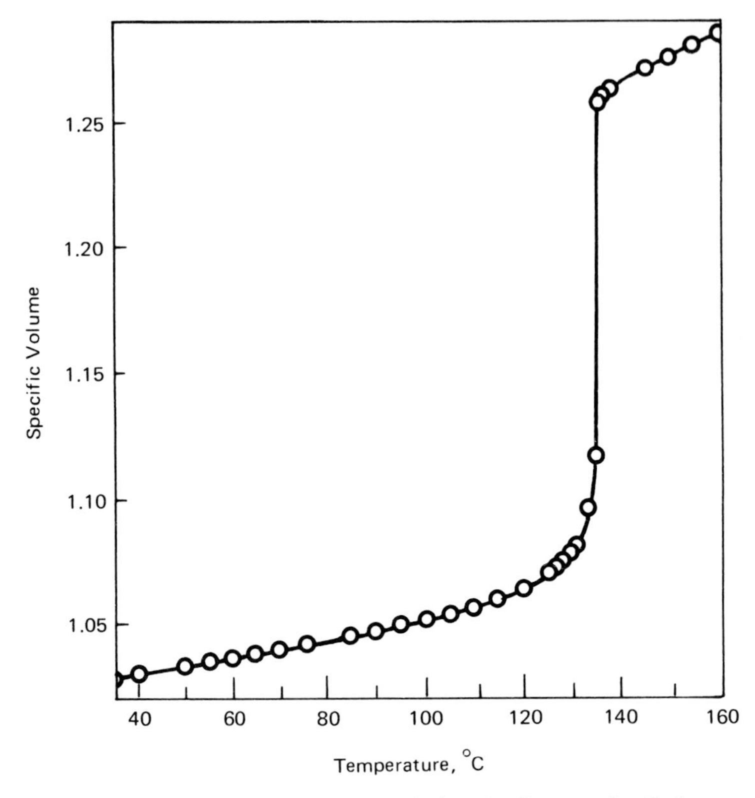

Figure 5.13. Specific volume-temperature relation for linear polyethylene, molecular weight 32,000. Adapted from R. Chiang and P. J. Flory, Journal of the American Chemical Society *83*, 2857 (1961). Reproduced with permission.

by the sharp increase in the specific volume in the vicinity of 136°C. The temperature at which the last traces of crystallinity disappear is well defined and is the melting temperature, which is designated as T_m. The fusion curve illustrated is comparable in all respects to those obtained with monomeric substances. There is a characteristic discontinuity in the specific volume or other comparable properties, at a well-defined temperature. Below the melting temperature properties of the crystalline state are observed. Above this temperature, depending on the molecular weight, the polymer will either be a viscous fluid or a rubber elastic body.

The sharpness of the fusion process illustrated in Figure 5.13 and its similarity to the melting of low molecular weight substances is a consequence of utilizing a molecular weight fraction and employing a very careful thermal treatment for crystallization and melting. A much more diffuse fusion process is observed for a crystalline polymer which is polydisperse in molecular weight and which is heated very rapidly. The melting range will now extend over many degrees, and the melting temperature is not nearly as well defined as in Figure 5.13. In the early studies of polymers, before their molecular constitution was recognized, the aforementioned kind of fusion curve was usually observed. Since it is almost a dogma in chemistry that pure substances must melt sharply, it was concluded that these materials were not pure.

It was further argued that since they were not pure substances they should not be studied. Consequently, they were discarded. In retrospect it has become abundantly clear that these were indeed chemically pure species. The broad melting range can be properly attributed to morphological rather than chemical impurities. The small crystallite sizes, their distribution, and the influence of the diffuse interfacial zone all contribute to the spreading out of the melting process. Superficially, the results are the same as if chemical impurities were present. We thus have a prime example of where the study of polymers has lead to the generalization of a widely held principle whose previous use had been highly restrained.

In contrast to the classical melting of homopolymers, the deliberate introduction of copolymeric units into a homopolymer chain influences melting in a very definite and systematic manner. We encounter here a phenomenon unique to long-chain molecules. The changes in the melting behavior depend not only on the amount and type of co-units introduced but also on how they are distributed along the chain. For example, if the co-units are randomly distributed, then a systematic depression of the melting temperature is found and the fusion process itself becomes very broad. However, if the co-units are introduced in long sequences, forming a block copolymer, then the melting temperature is invariant with compositional changes and the fusion process remains quite sharp. Figure 5.14 schematically illustrates the difference in the dependence of the melting temperature on composition between random and block copolymers. The melting temperatures of the corresponding pure homopolymers are T_{ma} and T_{mb} respectively. For the block copolymers the melting temperature, depending on composition, is identical to one of the pure homopolymeric species. For the random copolymers the melting temperatures are depressed from the pure species and reach a minimum value at a particular composition. The use of copolymers thus presents a very potent tool with which to control chemical and physical properties of crystalline polymers. If, for example, chemically more reactive groups are desired in a chain while it is necessary to maintain the melting temperature at the high value of the homopolymer, the co-units will be incorporated in an ordered manner. It may, however, be desirable to reduce the melting temperature for processing purposes. The random introduction of co-units does this. This procedure is commonly used in technological processes when very high melting polymers are involved.

Other kinds of chain irregularities such as geometric or stereo-isomers influence the melting temperature in a similar way. Thus, for example, while either highly isotactic or syndiotactic polypropylene forms a crystalline solid, when the stereo-isomers are randomly distributed along the chains a viscous liquid results. Although not always recognized, the presence of branched units also influences the melting behavior. For example, branching is characteristic of low density (as opposed to linear) polyethylene. This is a polymer that we have become very accustomed to in our everyday life. A comparison of the melting of branched and linear polyethylene is given in Figure 5.15. Although chemically identical, the melting point of the branched polymer is about 20 to 30 degrees lower than its linear counterpart. More striking, and more important from a practical point of view, is the much lower level of crystallinity characteristic of the branched polymer. Fusion takes place over a very wide temperature range, starting almost at room temperature, as compared

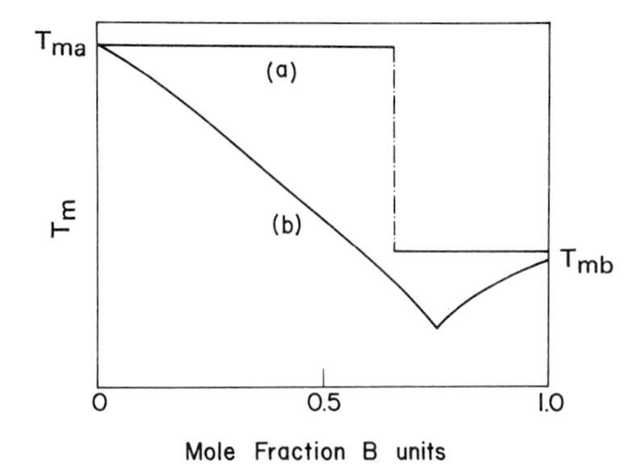

Figure 5.14. Schematic representation of the melting temperature-composition relation for copolymers. (a) Block copolymer. (b) Random copolymer.

to the very sharp melting of the linear polymer. These differences are naturally reflected in different physical properties and thus to the uses of these polymers.

Since the melting temperature of a homopolymer is a well-defined and important quantity and is easily determined, it is of interest to see how it is influenced by the chemical nature of the repeating unit. Table 5.1 presents a summary of the melting temperatures of some representative polymers. These data clearly indicate that the melting temperatures vary widely with chemical type. In this respect we find no difference between polymeric and monomeric substances. There are also many examples in the table that show how the location of the melting temperature governs the end-use of the polymer. For example, polymers commonly used as elastomers must have low glass temperatures. If they possess a regular chain structure, they must also have a low melting temperature. Otherwise the conformational requirements necessary for the elastic behavior will not be satisfied. We therefore find that three widely used elastomers–natural rubber, polyisobutylene, and polydimethyl siloxane–posses melting temperatures of 28°, 5°, and -40° respectively.[1] Thus, at room temperature they can be employed as rubbers. At the other extreme, polymers used as fibers must have very high melting temperatures. Two polymers extensively used as synthetic fibers, polyhexamethylene adipamide and polyethylene terepthalate, have very high melting temperatures, as does polyacrylonitrile. The para aromatic polyamides have very high melting temperatures as a consequence of their highly extended molecular conformation. The derivatives of cellulose are all very high melting. Cellulose trinitrate, used in gunpowder and in solid rocket propellants, has the highest known melting temperature of a polymer.

Polymers commonly used as plastics–such as polyethylene, polypropylene, and polymethylene oxide–have melting temperatures between those of fibers and elastomers. These polymers can be processed and fabricated in the molten state. When actually used they are in the crystalline state, so that the desired physical properties can be exploited, Such polymers are also called thermoplastics. The replacement of all the hydrogen atoms from linear polyethylene with fluorine atoms, to give poly-

[1] Although the melting temperature of natural rubber appears to contradict the principles just stated, it supercools very easily. Hence it is in the disordered, amorphous state at room temperature.

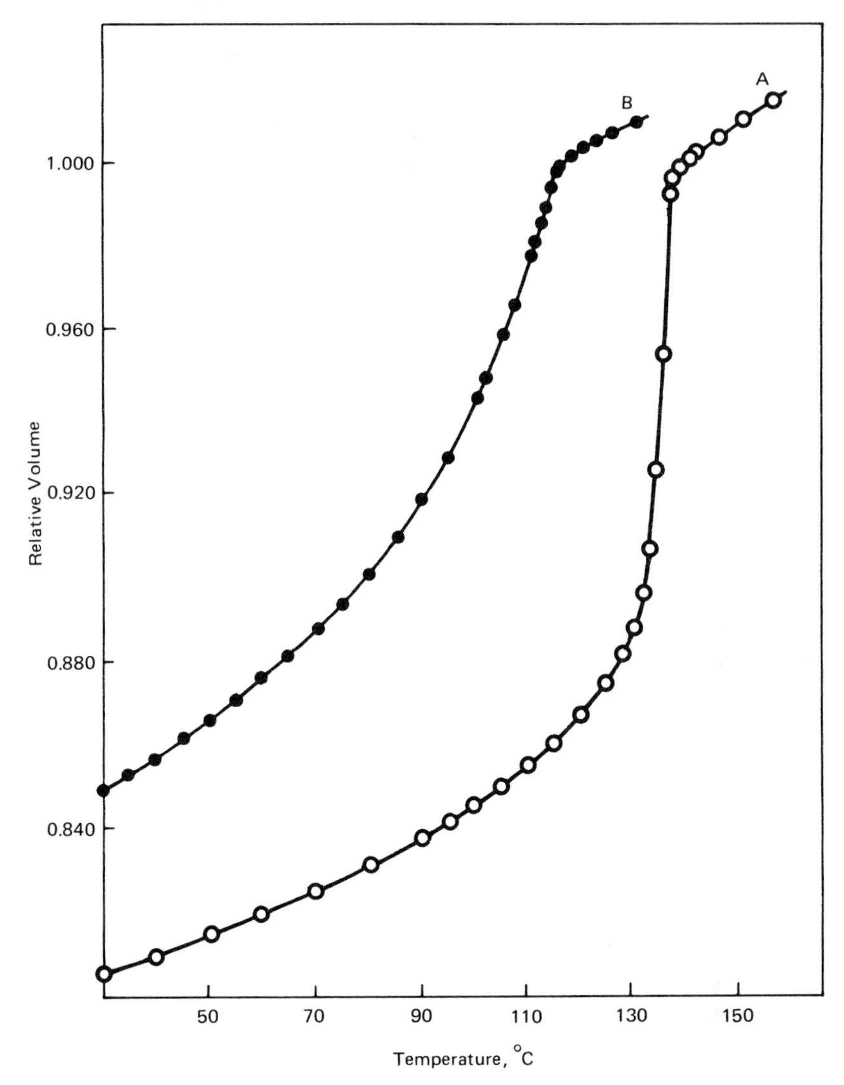

Figure 5.15. Plot of relative volume against the temperature for linear polyethylene, curve A, and branched polyethylene curve B. From Journal of the American Society *75*, 4093 (1953). Reproduced with permission.

tetrafluoroethylene, increases the melting temperature by about 200°C. This change in chemical constitution yields one of the most thermally stable polymers known. Poly(phenylene oxide), the polycarbonate polymers and poly(phenylene sulfide) are engineering plastics. They not only have high glass temperatures but also possess relatively high melting temperatures.

A detailed analysis of the molecular factors that determine the melting temperature has led to a causal relation with the disordered chain conformation. Generally speaking, the more flexible compact chains have the lower melting temperatures. The more rigid extended chains possess much higher melting temperatures.

Table 5.1. Melting Temperatures of Homopolymers

		T_m °C
Polydimethyl siloxane	Silicone rubber	-40
Polyisoprene–1,4 *cis*	Natural rubber	28
Polyisoprene–1,4 *trans*	Gutta-percha	75
Polyisobutylene	Butyl rubber	5
Polyethylene–linear		145
Polyethylene–branched		115-120
Polytetraflouroethylene	Teflon	330
Polypropylene		176
Polyacrylonitrile		317
Polymethylene oxide	Delrin	180
Polyhexamethylene adipamide	Nylon 6,6	270
Polyethylene terephthalate	Dacron	260
p-phenylene Polyamide	aramid, Kevlar	>500
Polydimethyl phenylene oxide		237
Poly bis-phenol-A-carbonate	Lexan	267
Polyphenylene sulfide		288
Cellulose triacetate	Acetate rayon	297
Cellulose trinitrate		>700

In Figure 5.16 we compare the mechanical properties, in the form of a stress-strain curve, between an initially unoriented crystalline polymer and a completely amorphous (lightly cross-linked) one. The initial slope, or modulus, is much less in the curve representing the rubber-like polymers. This observation reflects the fact that it is more easily deformed as a consequence of the differences in chain structure. At higher extensions there is a dramatic upsweep in the stress-strain curve caused by stress-induced crystallization. The chains have now become highly ordered and highly oriented in the stress direction leading to a very high modulus of elasticity. The system can no longer be easily extended or deformed, and rupture soon results.

Initially, the undeformed crystalline polymer has a very high modulus of elasticity and the stress-strain curve follows Hookes Law very closely at small elongations. However, at a relatively low elongation the crystalline polymer is said to yield (the maximum in the curve), and the crystallites and the chains become oriented in the chain direction. This drawing process results in relatively large extensions with a relatively small change in the force. At sufficiently high extension, failure ensues. Thus, the mechanical behavior of the two types of systems are drastically different up to the point of failure. These differences are a remarkable manifestation of the influence of molecular structure on mechanical properties.

Fibers

Fibers can be thought of as a special class of crystalline polymers. The crystalline regions in fibers are arranged so that the molecular chain axes are preferentially aligned, or oriented, in a direction parallel to the macroscopic fiber axis. This struc-

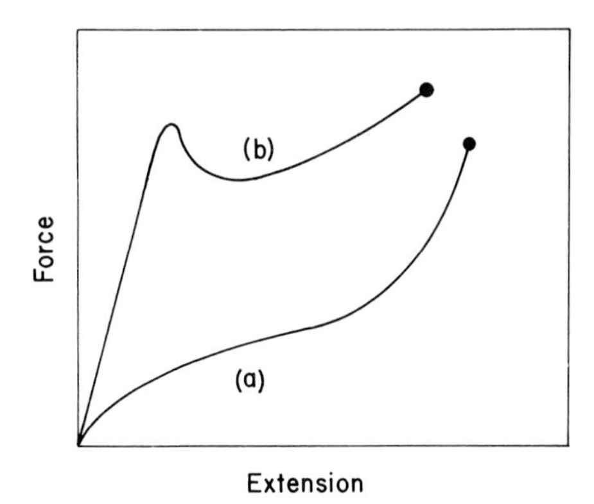

Figure 5.16. Comparison of mechanical properties between crystalline (unoriented) and amorphous rubber-like polymers. (a) Rubber-like polymer. (b) Crystalline polymers.

tural feature is common to all fibers irrespective of whether they are synthetic or "man-made," as nylon and dacron, or of natural origin as wool and silk. The outstanding physical properties of a fiber are its very high tensile strength, i.e., the ability to sustain large forces without breaking, coupled with a low deformability. As we would expect from our previous discussion, fiber properties are quite different from that of a rubber-like material and arise because of the major differences in chain structure and molecular organization.

There are two different methods by which synthetic fibers are in general formed. The distinction between the two depends on the chain conformation in the initial state prior to fiber formation. In the classical, or more conventional case, the chains are initally in random conformation. In more recent developments fibers are formed from polymers where the chains are initially in the liquid-crystal state. Here the individual molecules are in ordered conformation and are thus geometrically very highly assymmetric. This latter process might very well mimic the mechanism for fiber (fibril) formation in natural systems. In the conventional and widely used procedure, a bulk synthetic polymer, usually in the form of a powder, is converted to a fiber by a technological process known as spinning. The first step in spinning requires either the melting or the dissolution of the polymer. Polymers that can be melted without molecular degradation are melt-spun. Common fiber-forming polymers in this category are polyhexamethylene adipamide (nylon), polyethylene terepthalate (Dacron), and polypropylene. Because they are susceptible to thermal degradation, cellulose, cellulose acetate, and polymers based on polyacrylonitrile are solution-spun since the temperatures required are much lower than for melt-spinning. The principles involved in these spinning processes are very similar. The common factor here is that in either case the chains are in disordered or random conformation. We therefore will focus attention only on melt-spinning.

A schematic diagram of the apparatus involved in melt-spinning is given in Figure 5.17. Polymer, in the form of powder or chips, is fed from a hopper into the inert atmosphere of an electrically heated chamber maintained at a temperature above the melting temperature. The disordered polymer, in the liquid state, is then pumped

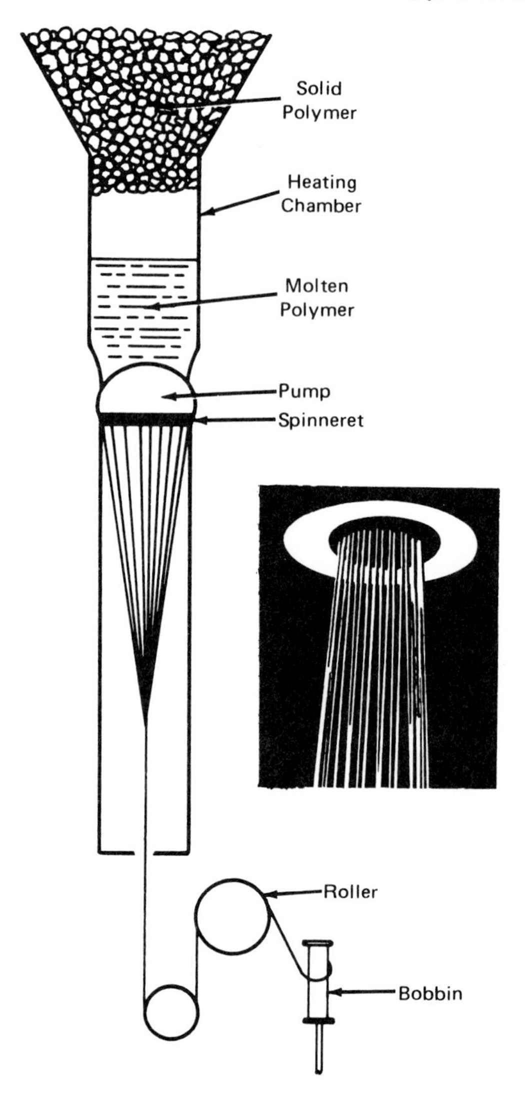

Figure 5.17. Schematic diagram of melt spinning process.

through a large number of very small holes that are drilled into a flat plate. This plate, called a spinneret, is shown in the inset of Figure 5.17. The liquid polymer emerges from the spinneret, in the form of thin threads, into air at a temperature well below the melting temperature. Crystallization naturally will ensue. At this stage in the process, however, the crystallite orientation is very poorly developed. The necessary high orientation for fiber properties is achieved by stretching, or "drawing," the crystalline threads. This is accomplished by passing the fibers through two pairs of rollers rotating at different speeds. The second roller is operated four to five times faster than the first one. This arrangement yields a net tensile force that stretches the fiber several hundred percent. During drawing, the temperature between the two sets of rollers is kept above the glass temperature but below the melting temperature of the polymer. The spinning process yields a very high crystallite orientation in a fibrous polymer. It does so in a very effective manner, since a large quantity of fiber can be continuously produced.

The high state of crystallite orientation is evidenced by wide angle X-ray patterns similar to those illustrated in Figure 5.4(c). However, despite the great ingenuity that has been applied and the mechanical devices that have been developed for orienting crystalline polymers, the extent and perfection of the orientation that can be developed by this type process is not yet as great as is found in the naturally occurring fibers such as silk, wool, and cotton. In synthetic fibers the lamella-like crystallite structure must obviously have been reduced to some extent. The details of the morphological structure are still being debated. It is clear, however, that fibers formed in this manner must consist of a periodic repetition of crystalline and amorphous zones. These structural features must of course turn up in the mechanical properties of the fibers.

In the liquid-crystalline state, as illustrated in Figure 3.13, the molecules are already partially oriented parallel to one another as in a crystal. This situation results because of the high assymmetry of the individual ordered chains and their inability to exist in the isotropic state, except at very high dilution. The processing of liquid-crystalline systems by methods similar to those described in Figure 5.17 leads to fibers of very high orientation. In this case it is not necessary to deform, orient, and crystallize the chains. High extension and partial axial orientation are already built into the system, so to speak. Subsequent processing merely enhances the ordering and leads to very high axial orientation comparable to that found in natural occurring fibrous systems. The very high orientation that can be developed by spinning from the liquid-crystal state has a very profound effect on mechanical properties and ultimate strength of the fibers formed.

A typical stress-strain curve for a synthetic fiber is given in Figure 5.18. These curves are quite different from that of rubber. The long-range elasticity is clearly lost. The maximum extension being relatively small, on the order of 2 to 5%. The forces required for these extensions are, however, very large. The initial portions of the curves, representing the first few percent of extension, are a result of the deformation of the noncrystalline chain units. Further deformation takes place at the expense of more orientation and eventually the destruction of the crystalline regions. This process leads to a very steep stress-strain curve. The high stresses involved eventually lead to chain rupture and fracture of the sample. The important physical properties of fibers are high strength accompanied by a high modulus of elasticity, i.e., resistance to deformation. These properties are directly related to and are a consequence of the chain structure and morphology. The highly ordered arrangements of the chain atoms with the accompanying loss of conformational freedom yields a high modulus of elasticity and high strength as compared to rubber-like polymers. The very enhanced axial orientation of the aromatic polyamides relative to the more conventionally formed fibers of poly(ethylene terepthlalate)—dacron—and the aliphatic polyamides (nylons) results in a 10- to 20-fold increase in the initial modulus, i.e., they are 10 to 20 times stiffer. Furthermore, the elongation, or extension, at break is only about 2% for these fibers.

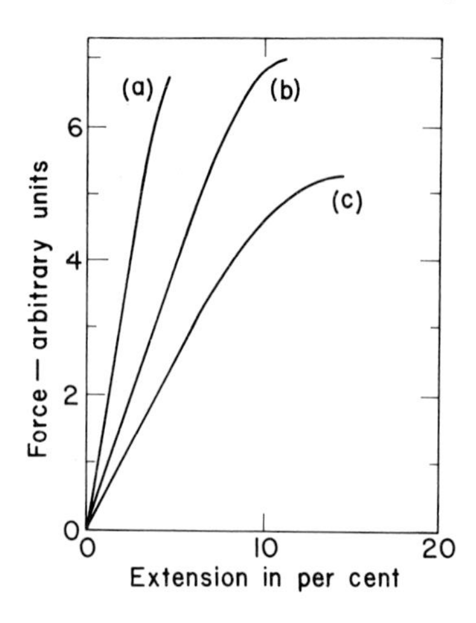

Figure 5.18. Force-length curves for synthetic fibers. (a) Aromatic polyamide. (b) Polyethylene terephtalate (dacron). (c) Aliphatic polyamide (nylon).

These fundamental fiber properties are not directly derived from any particular chemical structure. They basically result from the typical state of high order and orientation that always exists. However, for a fiber to be useful as a textile material other conditions also have to be satisfied. For example, to be able to undergo ironing and washing a fiber-forming polymer must have a high melting temperature so that it remains intact during these processes. Consequently, as we can tell from the data in Table 5.1, melting temperatures of at least 200°C are necessary to form a satisfactory fiber. Thus, although highly ordered and oriented fibers of linear polyethylene can be prepared, their use is fairly limited because of the relatively low melting temperature of 145°C. On the other hand, the aromatic polyamide fibers, the aramids, maintain their strength and other desirable mechanical properties to temperatures in excess of 300°C because of their high melting temperatures. The melting temperature is of course determined by the chemical nature of the repeating unit. To be attractive a fiber must be capable of accepting various kinds of dyes. This condition also limits the types of chemical repeating units that can be used. Copolymeric units are very often introduced to improve the dyeing capability.

It seems appropriate at this point to summarize some of the mechanical properties of polymers and to review how they depend on the different states. The stiffness or deformability of a substance can be quantitatively represented by its Youngs modulus, which is the ratio of the applied stress (force/unit area) to the relative extension (strain). The chart of Figure 5.19 gives some typical values for polymers and monomers. The relative ease of deformation is inversely related to the Youngs modulus. An enormous range in properties is represented in Figure 5.19 even if considerations are restricted to polymeric substances. Crystalline, nonpolymeric materials, such as diamond and metals, represent the least deformable substances known. They are closely followed by inorganic glasses whose moduli lie in the range 10^{11} to 10^{12} dynes/cm^2. The moduli of polymeric fibers are in the lower portion

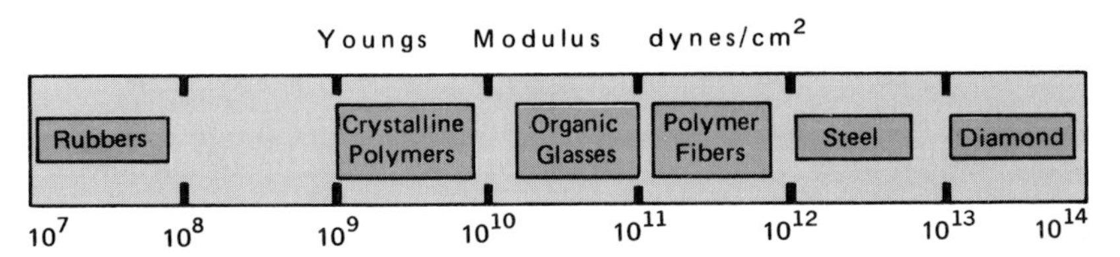

Figure 5.19. Youngs modulus for different substances.

of this range. The value for nylon is the order of 10^{10}-10^{11} dynes/cm^2. As we might expect, the modulus for *p*-phenylene polyamide is much greater. Depending on the particular sample, it ranges from 1.1 to 1.4×10^{12} dynes/cm^2. The moduli progressively decrease for organic glasses, unoriented crystalline polymers, followed by rubbers, which have the lowest value. The magnitude of the modulus must be intimately related to the molecular mechanisms responsible for the elastic response. This in turn in dependent on the molecular and morphological structures present. From our knowledge of polymer conformations and of the different crystallite arrangements that can exist, the large variation in the modulus should not surprise us. The inelasticity of diamond and similar materials results from the fact that the constituent atoms are very firmly and rigidly bound to one another in a crystal lattice and can therefore not be easily displaced.

Another physical property of great practical importance is tensile strength, the ultimate load that a material can bear without breaking. Some typical values of this quantity are given in Table 5.2. The strength of the materials order themselves in the manner that we would expect. The range in values, however, is not nearly as great as for the elastic modulus. The strongest materials are usually the least deformable. Therefore, one must always seek a compromise between high strength and high elasticity. The strength of a material is directly related to the molecular order and crystallite orientation. Crystalline, but unoriented nylon is only slightly stronger than rubber. However, when the crystallites are oriented, as in a fiber, the tensile strength increases more than tenfold. Less oriented fibers such as cotton and viscose rayon have a somewhat lower tensile strength. The ulitmate strength data, given in

Table 5.2. Tensile Strength of Materials

Materials	Tensile Strength *dynes/cm*^{2}
Steel, piano wire	2×10^{10}
p-phenyl Polyamide, aramid fiber	2×10^{10}
Nylon, fiber	1×10^9
Cotton, cellulose	5×10^9
Viscose rayon, textile fiber	3×10^9
Wool	2×10^9
Nylon, crystalline unoriented	8×10^8
Polymethyl methacrylate, glass	8×10^8
Rubber	3×10^8

Table 5.2, have been calculated on the basis of the cross-sectional area of the sample. On this scale the tensile strength of synthetic fiber is about half that of a high-strength steel wire. The values for steel and the aromatic polyamide are comparable. However, there is a large difference in the densities (mass/unit volume) of the materials. Steel is almost seven times denser than nylon, and about five times more dense than the aromatic fiber. Therefore, when tensile properties are compared on a mass basis the nylon fiber is about three-and-one-half times stronger than the steel wire, and the aramid fiber about five times. These facts highlight one of the most important properties of polymeric materials. Namely, that when compared on a mass basis they are very strong indeed. The high strength of polymeric fibers can be illustrated by the fact that a 1-inch diameter nylon rope can easily lift a 5½-ton load. Thus, such fibers are both ultra-light and ultra-strong.

Mechanochemistry

In addition to the desirable mechanical and physical properties that make polymeric fibers so useful, they also possess another very important characteristic. This is a dimensional property that allows all fibrous macromolecules to be contractile, i.e., to undergo anisotropic dimensional changes. The fundamental reason for this behavior is the molecular dimensions that are associated with the distinctly different conformations of the crystalline state as compared to the liquid or disordered state. We recall that in the liquid state, where the chains are in statistical conformation, a linear dimension of a chain molecule is conveniently represented by the root-mean square distance between chain ends, $\langle r^2 \rangle^{1/2}$. This dimension is directly proportional to $n^{1/2}$, the square root of the number of bonds making up the chain. On the other hand, in the crystalline state, with an ordered chain conformation, a linear molecular dimension is proportional to n. For a high molecular weight chain, the ratio of these two dimensions can be quite large. For example, for a chain comprised of 10^4 bonds, a 100-fold decrease in molecular length will occur upon disordering.

The question then arsies as to whether these molecular dimensional changes can be manifested in a gross macroscopic change during the transformation from one state to the other. Again a distinction must be made with respect to the nature of the crystalline state. If the crysatlline regions are randomly arranged relative to one another, the local dimensional changes will average out. Therefore, the expected macroscopic changes will not occur. However, when the molecular axes of the ordered chains are preferentially oriented along a macroscopic dimension in space, as in a fiber, then the expected dimensional changes are indeed observed. The magnitude of this change will depend on the extent and perfection of the orientation. This dimensional property is what causes the shrinkage of wool and other synthetic fibers. In the laboratory this phenomenon can be studied more quantitatively. In Figures 5.20 and 5.21 length-temperature relations are given for two very different polymers, fibrous natural rubber and the fibrous protein collagen. Although these fibers are very different chemically and their use is quite different, structurally they are both crystalline and axially oriented. Consequently, upon melting a large diminution in length is observed for both polymers. This change occurs very suddenly over a very small temperature interval. The plots in Figures 5.20 and 5.21 have similar characteristics to the volume-temperature plot of Figure 5.13. In the latter case the volume increases over a very narrow temperature interval. The razor-like sharpness of the change, whether it be in volume or in length, is a consequence of the fact that a melting transition is taking place.

The dimensional changes that accompany the melting process will be reversible if an equilibrium force is imposed on the system or if an adequate number of intermolecular cross-links are introduced into the original fibrous structure. In Figure 5.22 we have an example of a reversible, contractile polyethylene fiber. There is no external force acting on the fiber but it has been cross-linked. On the heating cycle shortening is observed concomitant with melting; on cooling and recrystallization re-elongation to the original dimensions takes place.

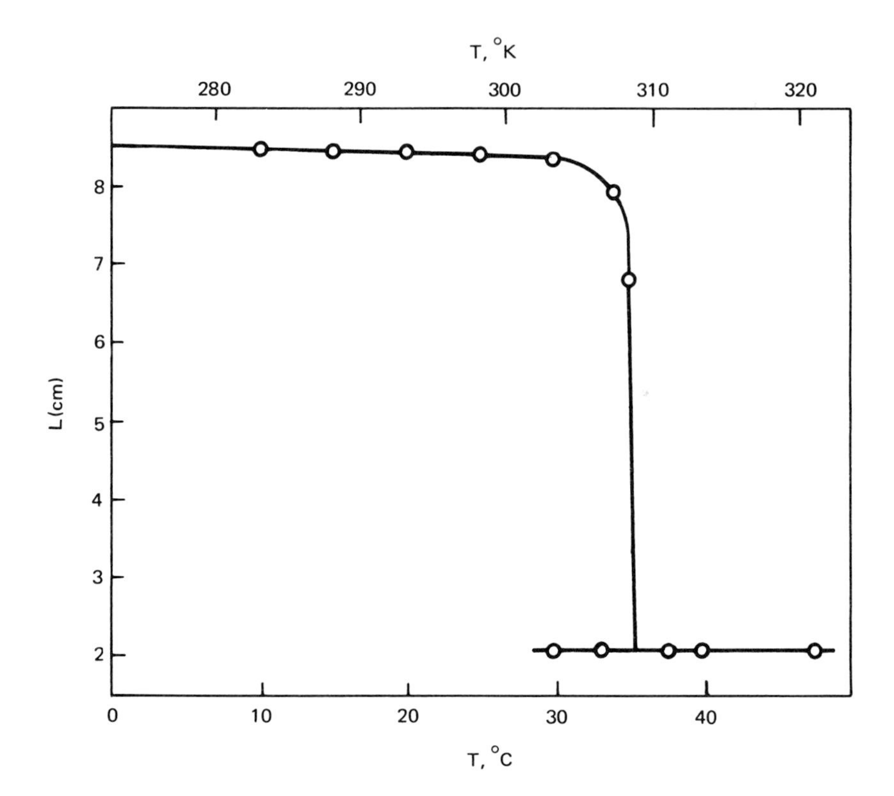

Figure 5.20. Length-temperature relation for fibrous natural rubber under zero force. From J. F. M. Oth and P. J. Flory, Journal of the American Chemical Society *80*, 1297 (1958). Reproduced with permission.

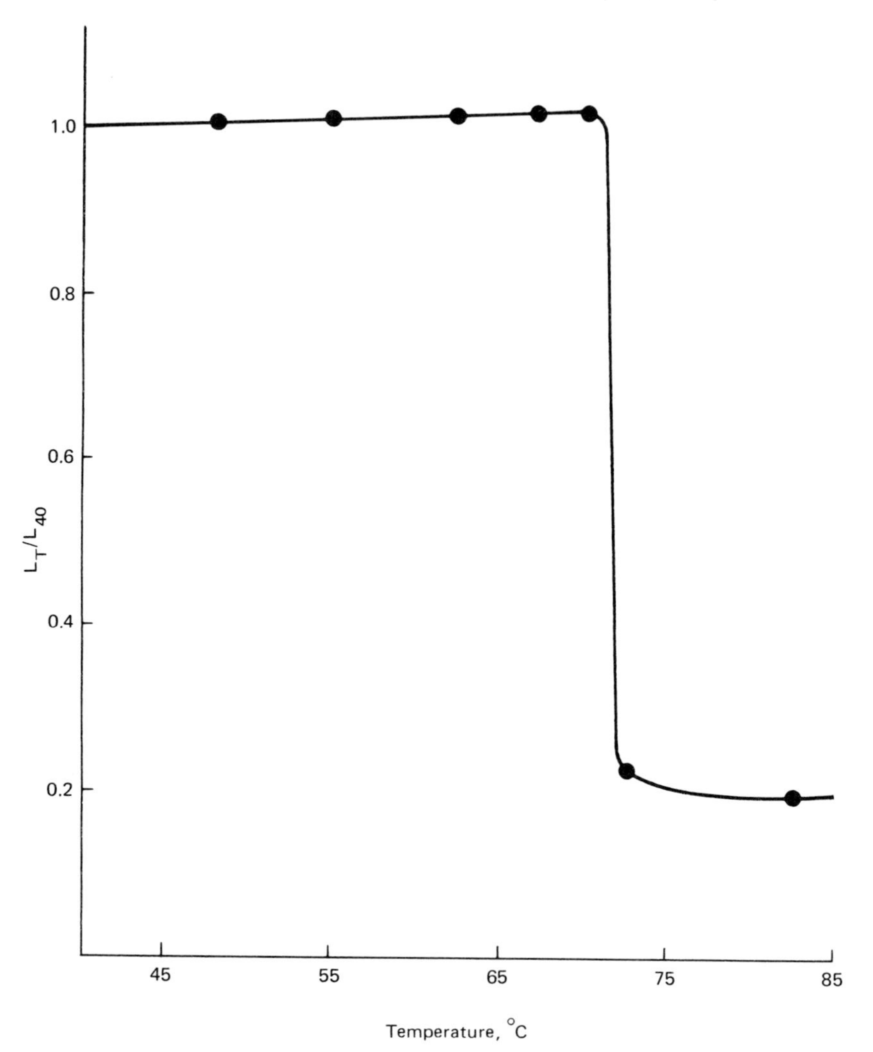

Figure 5.21. Relative length-temperature relation for rat tail collagen under zero force. From P. J. Flory, Journal Cellular and Comparative Physiology *49* (Supplement 1), 175 (1957). Reproduced with permission.

Figure 5.23 illustrates the length-temperature relations that will be observed under a constant force. The melting temperature will increase with an increase in the applied force.[2] Shrinkage occurs over a very narrow temperature range but the magnitude of the change will decrease with an increase in the force. With this information we can now consider another problem: if the length is to be held constant, what force must be imposed, at a given temperature, to satisfy this condition? The answer is that in order to prevent fusion and shrinkage the force must just exceed

[2] This is completely analogous to the increase in the melting temperature that is observed for virtually all substances, with an increase in the applied hydrostatic pressure.

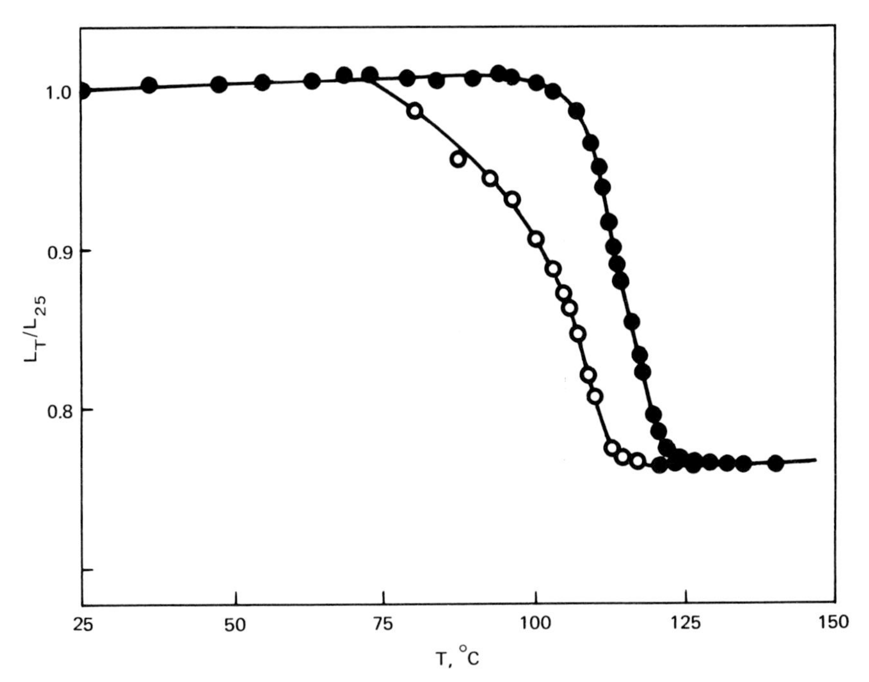

Figure 5.22. Relative length-temperature relation for reversible contractile polyethylene fiber. ● heating; ○ cooling. From Journal of the American Chemical Society *81*, 4148 (1959). Reproduced with permission.

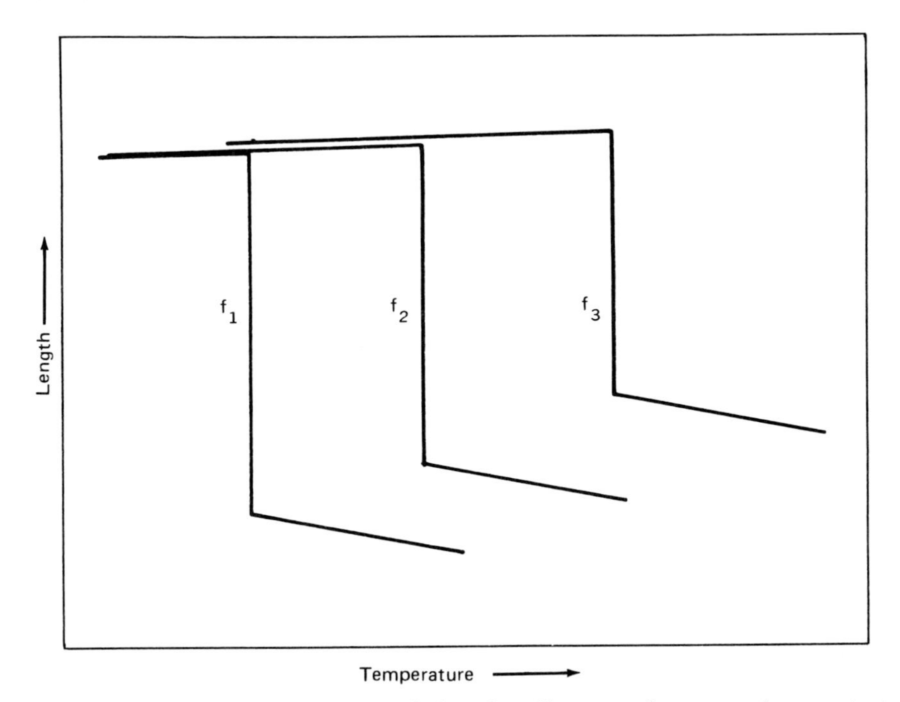

Figure 5.23. Length-temperature relation for fibrous polymers under constant force f. $f_3 > f_2 > f_1$. Adapted from P. J. Flory, Science *124*, 53 (1956). Reproduced with permission.

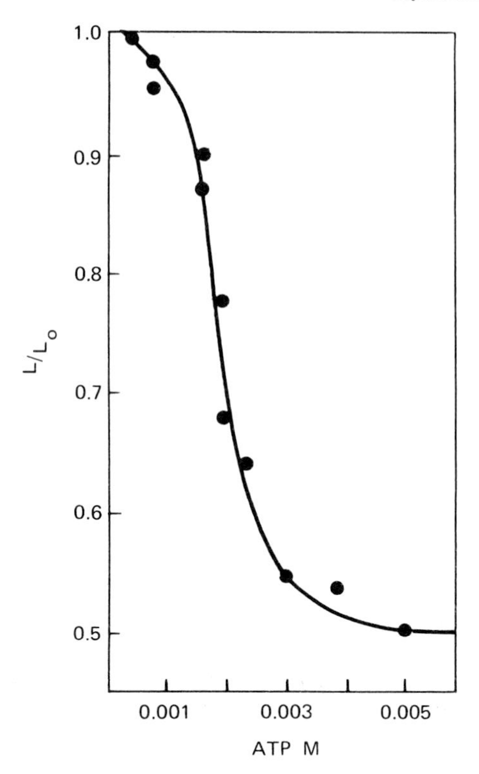

Figure 5.24. Relative length change of glycerinated muscle fibers as ATP concentration is varied. From Proceedings of the National Academy of Sciences *45,* 814 (1959). Reproduced with permission.

the equilibrium one. This force is known as the isometric force and is equal but oppositely directed to the force developed by the fiber upon melting. For synthetic fibers, isometric tensions as high as 4 kg/cm^2 have been observed. This tension is of the same order of magnitude as is developed by natural muscle systems in tetanic contraction.

In the examples illustrated, melting and the accompanying dimensional changes, or tension development, were caused by changes in temperature. Identical processes can, however, be carried out at constant temperature. They require more reactive fibers that can chemically interact with appropriate reagents. Isothermal chemical melting, accompanied by corresponding changes in dimensions and in tension, has been amply demonstrated with many of the fibrous proteins. We give, as an example in Figure 5.24, the shortening of glycerinated muscle fiber, a conventional muscle model, when immersed in a solution of adenosine triphosphate (ATP). In complete analogy with the previous figures, a major diminution in length occurs with a small change in ATP concentration. This dimensional change is accompanied by the loss of the ordered fibrous structure. When experiments are conducted at constant length, large tensions develop at the same ATP concentration that the major shortening occurs in Figure 5.24. There is, therefore, a very simple and general process, applicable to all fibrous macromolecules, that allows for anisotropic dimensional changes and tension development on a macroscopic level. The only structural requirement is that the system be fibrous.

Contractility and tension development are very important phenomena that are observed in many natural occurring processes such as muscular contraction, cell motility, and cell mitosis. These processes invariably involve the utilization of macromolecules.

Although the laboratory experiments described in Figure 5.24 involve species of physiological interest, the question as to whether the mechanism outlined above is indeed operative in natural functioning systems is the subject of intensive investigations in many laboratories. The biologically interesting systems usually have very intricate structures and are governed by complex chemical reactions.

Chapter 6

MACROMOLECULES OF BIOLOGICAL IMPORTANCE

Introduction

In this chapter we shall discuss the structure and properties of macromolecules of biological and biochemical importance, primarily proteins and nucleic acids. Proteins can be thought of as naturally occuring copolymers of L-amino acids. In contrast to the usual synthetic copolymers, however, the sequential arraangement of the repeating units is very precisely and exactly defined. The particular arrangement leads in turn to a very definite structure that controls the function of the protein.

Proteins, which are of prime importance to life (the word "protein" comes from the Greek *proteios* meaning first rank), serve in many diverse roles. For example, they form the structural materials that hold living organisms together, they transport ionic and molecular species within a living system, they control and catalyze virtually all metabolic processes, and as antibodies they combat viral and bacterial diseases. A staggering number of different proteins is found in all forms of animal life.

Such great versatility in the behavior of a given type of macromolecule may at first be surprising. However, 20 different amino acids, or co-units, are found in proteins. Hence, even for a chain containing only 100 repeating units, it is possible to have 20^{100}, or 10^{300}, chemically distinguishable molecules. When the conformational versatility of the chain is coupled with these vast chemical possibilities, the great diversity of structure becomes apparent. In fact, only a very small fraction of the theoretical number of distinct molecules is actually found in nature. Before engaging in a more detailed discussion of protein structure and function, it is instructive to examine the properties of homopolymers formed from the L-amino acids. These polymers are called polypeptides.

Polypeptides

The repeating unit of a polypeptide chain is an amino acid whose skeletal structure is given in Figure 6.1. In this monomer the α-carbon atom, C^{α}, is flanked by an amino group (NH_2) on one side and a carboxyl group (COOH) on the other. A hydrogen atom and an R group are attached to the α-carbon. The R group represents a particular substituent or side chain. Twenty different kinds of R groups, and

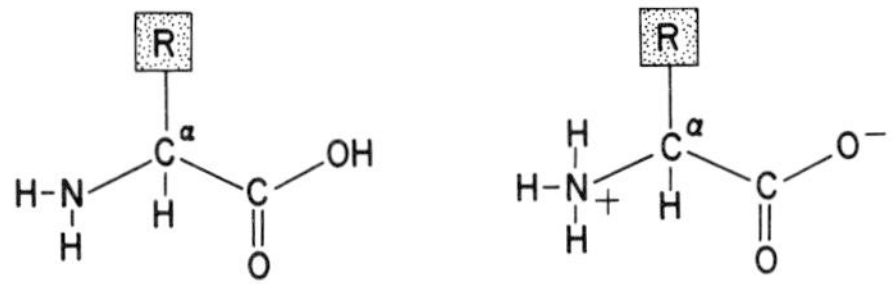

Figure 6.1. Amino acid structure.

thus amino acid repeating units, are found in proteins. In solution, the amino acid can exist as a doubly charged species or zwitterion. There is a positive charge on one end of the molecule and a negative one on the other.

Because there are four different groups attached to the α-carbon (except when R = H), the amino acids can exist in two stereo-isomeric forms that are mirror images of one another. The two isomers are shown in Figure 6.2. Of the two possible ways

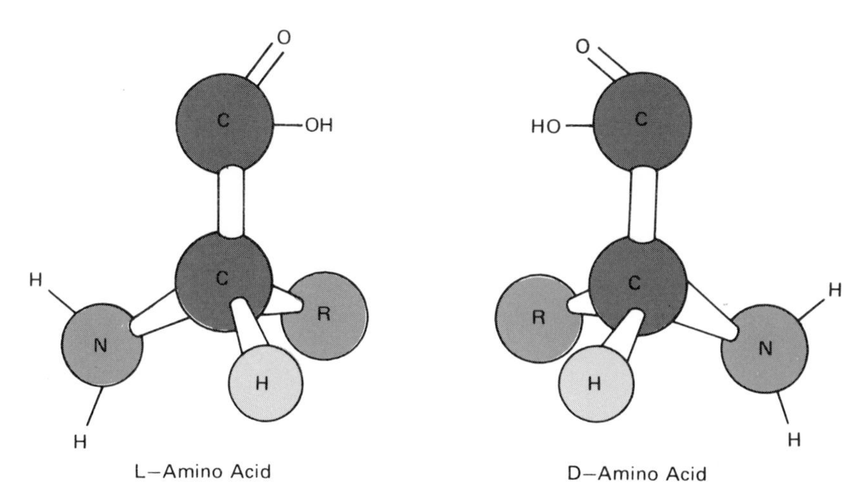

Figure 6.2. The two stereo-isomers of the amino acids.

that the groups can be attached to the α-carbon only one, the L isomer, is found in proteins. Although the other isomer is occasionally found in other naturally occurring molecules, it has not as yet been found in a protein. The reason for this strong preference for the L isomer is not as yet clear. The answer is clearly involved with the question of the origin of life itself.

Structural formulas for the 20 amino acids that constitute the repeating or co-units of proteins are listed in Table 6.1. Their formal names as well as the commonly accepted abbreviations are also given. The most striking feature of the table is the very wide range of chemical structures that are represented. These range from hydrocarbon type side chains (such as Gly, Ala, Val, and Leu) to those containing aromatic groups (such as Phe). Included are hydroxyl (OH)-containing residues (such as Ser and Thr) and charged acidic and basic groups (such as Asp and Glu and His, Lys and Arg). Proline and hydroxyproline are special, since the side groups are five-membered rings that include the α-carbon and adjacent nitrogen atoms. Cystine plays a unique role because the disulfide bond can act as a cross-linker between chains or between units of the same chain. In addition to the statistical feature of distributing the amino acid residues along a chain, the very distinct and different chemical properties are also important in determining structure and function. Since water is the natural habitat of protein molecules, we must recognize that the polar and charged groups show a strong desire to mix with water. These are termed hydrophilic groups. On the other hand, the hydrocarbon and nonpolar side groups would prefer to be removed from the aqueous medium. Such groups are called hydrophobic.

Table 6.1. Amino Acids Found in Proteins

Amino acid	Chemical Structure
Glycine (Gly)	$H-CH(NH_2)-COOH$
Alanine (Ala)	$CH_3-CH(NH_2)-COOH$
Valine (Val)	$(CH_3)_2CH-CH(NH_2)-COOH$
Leucine (Leu)	$(CH_3)_2CH-CH_2-CH(NH_2)-COOH$
Isoleucine (Ileu)	$CH_3-CH_2-CH(CH_3)-CH(NH_2)-COOH$
Serine (Ser)	$HO-CH_2-CH(NH_2)-COOH$
Threonine (Thr)	$CH_3-CH(OH)-CH(NH_2)-COOH$
Tyrosine (Tyr)	$HO-C_6H_4-CH_2-CH(NH_2)-COOH$
Phenylalanine (Phe)	$C_6H_5-CH_2-CH(NH_2)-COOH$
Tryptophan (Try)	$C_8H_5N(H)-CH_2-CH(NH_2)-COOH$ (indole ring: C–CH$_2$, =CH, N–H)
Aspartic (Asp)	$HOOC-CH_2-CH(NH_2)-COOH$

Table 6.1. (continued)

Amino Acid	Chemical Structure
Glutamic (Glu)	$HOOC{-}CH_2{-}CH_2{-}CH(NH_2){-}COOH$
Lysine (Lys)	$NH_2{-}CH_2{-}CH_2{-}CH_2{-}CH_2{-}CH(NH_2){-}COOH$
Arginine (Arg)	$NH_2{-}C({=}NH){-}NH{-}CH_2{-}CH_2{-}CH_2{-}CH(NH_2){-}COOH$
Histidine (His)	$HC{=}C(NH{-}CH{=}N){-}CH_2{-}CH(NH_2){-}COOH$ (imidazole ring: HC=C–NH–C(H)=N–)
Methionine (Met)	$CH_3{-}S{-}CH_2{-}CH_2{-}CH(NH_2){-}COOH$
Cysteine (CySH)	$HS{-}CH_2{-}CH(NH_2){-}COOH$
Cystine (Cys)	$HOOC{-}CH(NH_2){-}CH_2{-}S{-}S{-}CH_2{-}CH(NH_2){-}COOH$
Proline (Pro)	$CH_2{-}CH_2{-}CH_2{-}N(H){-}CH{-}COOH$ (ring)
Hydroxyproline (Hypro)	$HO{-}CH{-}CH_2{-}CH(COOH){-}N(H){-}CH_2$ (ring)

The formation of a polypeptide chain from the amino acid monomers is conceptually similar to the classical condensation polymerization. The monomer is bifunctional, and one group can react with the other while the bifunctionality remains intact in the resulting molecule. Hence, by the principles developed in Chapter 2, long-chain molecules can be formed as is shown in Figure 6.3. This reaction yields

```
      H O               H H O                       H O H H O
      | ||              | | ||                      | || | | ||
NH2-C-C- OH + H -N-C-C-OH  →  NH2-C-C-N-C-C-OH + H2O
      |                   |                         |       |
      R                   R                         R       R
```

Figure 6.3. Formation of a dipeptide from two amino acids.

a dipeptide with an amino and a carboxyl end group. Further stepwise reactions yield long-chain polypeptides with the same terminal groups as is shown in Figure 6.4. A large number of different copolymers can be devised depending on the chemical nature of the R groups and their sequential arrangement along the chain.

```
     H O H H O H H O H        H  OH
     |α || | |α || | |α || |        |α /
NH2-C-C-N-C-C-N-C-C-N·········C-C
     |       |       |              |  \\
     R       R       R              R   O
```

Figure 6.4. Formation of polypeptide chain.

Conformational Properties of Polypeptides

From the skeletal chain structure in Figure 6.4 we can see that there are three different bonds that span the distance from one α-carbon atom to the next one. The rotational characteristics of the C^{α}–C bond, the C–N bond, and the N–C^{α} bond are, therefore, very important in determining the conformational properties of the chain. The electronic structure of the C–N, or amide bond, is such that it possesses a great deal of double-bond character. In effect, this bond is nonrotatable. This characteristic leads to a planar amide group with the sequence of atoms from one α-carbon to the next, the grouping C^{α}–CO–NH–C^{α}, being constrained to lie in a plane of considerable rigidity such as is shown in Figure 6.5. However, rotations are still possible

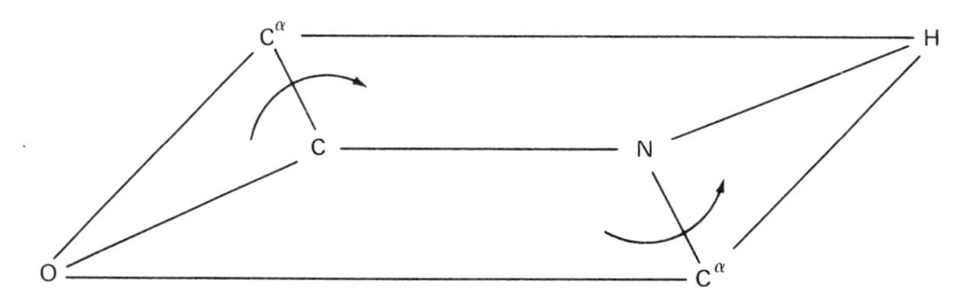

Figure 6.5. Planar amide group.

about the C^{α}–C and the N–C^{α} bonds without altering the required planarity. The polypeptide chain will thus still possess a great deal of conformational versatility. The α-carbon atoms can be thought of as junction points that allow for the orientation of one planar group relative to its neighbor.

Polypeptides can thus adopt statistical conformations in the same manner as other chain molecules. The characteristics ratio, $C = \frac{\langle r^2 \rangle}{nl^2}$ will be expected to depend on the chemical nature of the side group. For polyglycine (R = H), the characteristic ratio is 2. This low value indicates a very compact chain having considerable rotational freedom about the two single bonds. On the other hand, when bulkier side groups are present, such as in poly-L-alanine ($R = CH_3$), the characteristic ratio increases to about 9, indicating a more extended chain structure. The poly-L-proline chain represents a special situation as can be seen from the skeletal structure given in Figure 6.6. Here the R group is a five-membered ring, and one of the ring bonds

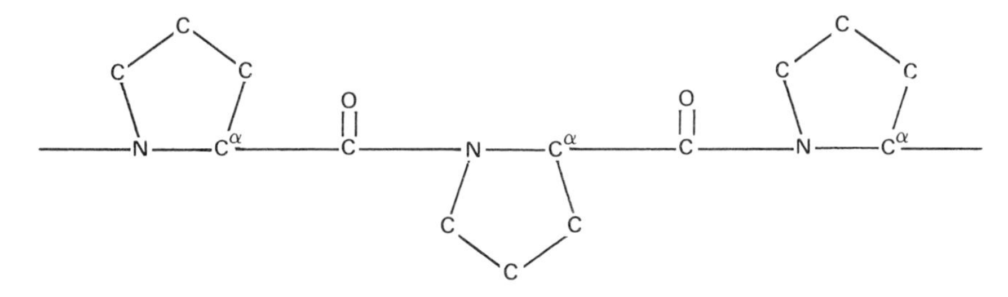

Figure 6.6. Skeletal structure of poly-L-proline.

is part of the main chain. The C–N bond still possesses its double-bond character and is essentially nonrotable. Since the N–C^{α} bond is part of the ring, its rotational angle is also fixed. Hence, for this chain, rotation is only allowed about the C^{α}–CO bond. This rotation will, however, be severely restricted because of the steric interactions of the two adjacent rings. A highly extended moelcule can be expected and indeed is found since the characteristic ratio for poly-L-proline is about 20.

When a preferred set of rotational angles are repeated successively along the chain, then an ordered conformation is generated, as we have found for other polymers. The ordered polypeptide structures can actually be observed in the isolated molecules in dilute solution. Polypeptides form three known major types of ordered structures.

The most widely known ordered structure is the α-helix of Pauling and Corey. Figure 6.7 (a) shows how a single polypeptide chain can be twisted is such a way as to form the helical structure of Figure 6.7 (b) while the planarity of the amide groups is maintained. Figure 6.7 (b) is a model of the right-handed α helix. The salient dimensional and geometric features of this conformation are shown in Figure 6.8. The chain backbone follows the pattern of the thread of a right-hand screw. An exact repeat of the structure occurs every 18 repeating units (or amino acid residues). This repeat corresponds to five turns of the helix or a linear advance of 27 Å. Each turn of the helix thus involves 3.6 residues, so that the helix is nonintegral. The recognition that nonintegral helices were permissible structures provided the

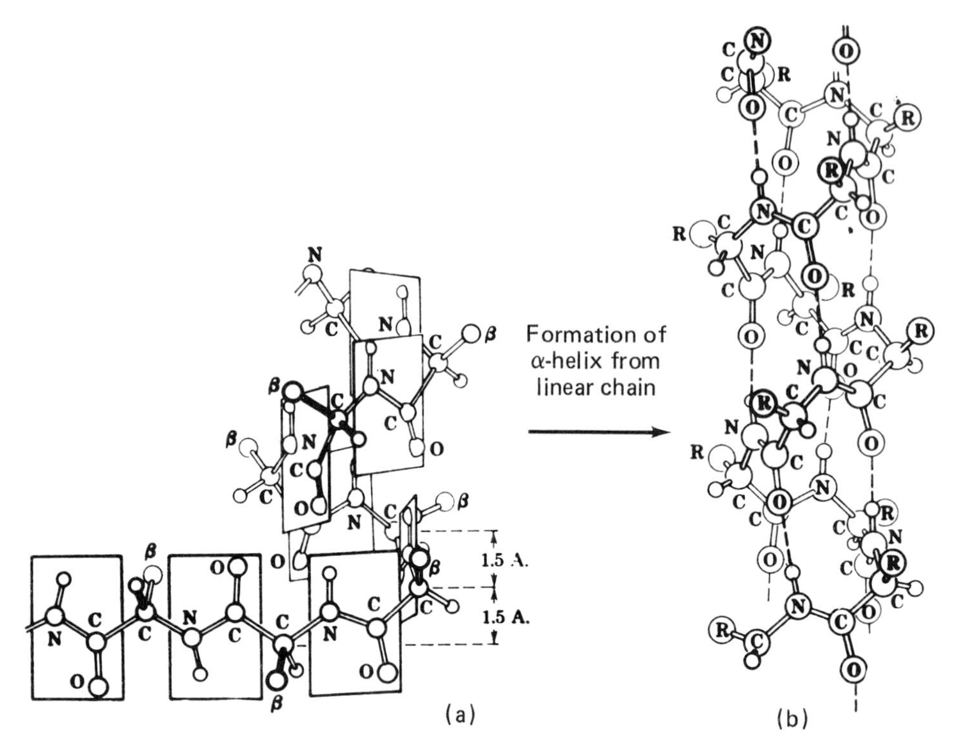

Figure 6.7. Schematic representation of twisting an extended polypeptide chain into the α-helical conformation.

necessary insight into solving this particular structural problem. This solution in turn opened the way for understanding the details of protein structure as well as the structure of other macromolecules pertinent to molecular biology.

A very important feature unique to the α-helix is indicated by the dashed lines of Figure 6.7 (b). The spatial positions of the atoms are such that the hydrogen atom from a N–H group forms a hydrogen bond with the oxgen atom of the C=O group on the third amino acid residue down the chain. In this structure the bonding atoms find themselves the proper distance apart on successive turns of the helix. This *intramolecular* hydrogen bond is a feature unique to the α-helical structure, giving it a very stable conformation. A well-defined, ordered structure can thus be formed and stabilized within a single polypeptide chain without any intermolecular interactions being involved.

The fact that polypeptide chains can form an ordered helical structure should not surprise us. We have already encountered this kind of structure in our study of synthetic polymers, particularly those that are stereo-regular and isotactic. From a purely geometric point of view, without concern for specific atomic positions and interactions, there is a great deal of structural similarity between the two types of chains as is seen in Figure 6.9, which compares the α-helix formed by polypeptides and the 3.5 helix (3.5 monomer units per turn) generated by an isotactic polymer.

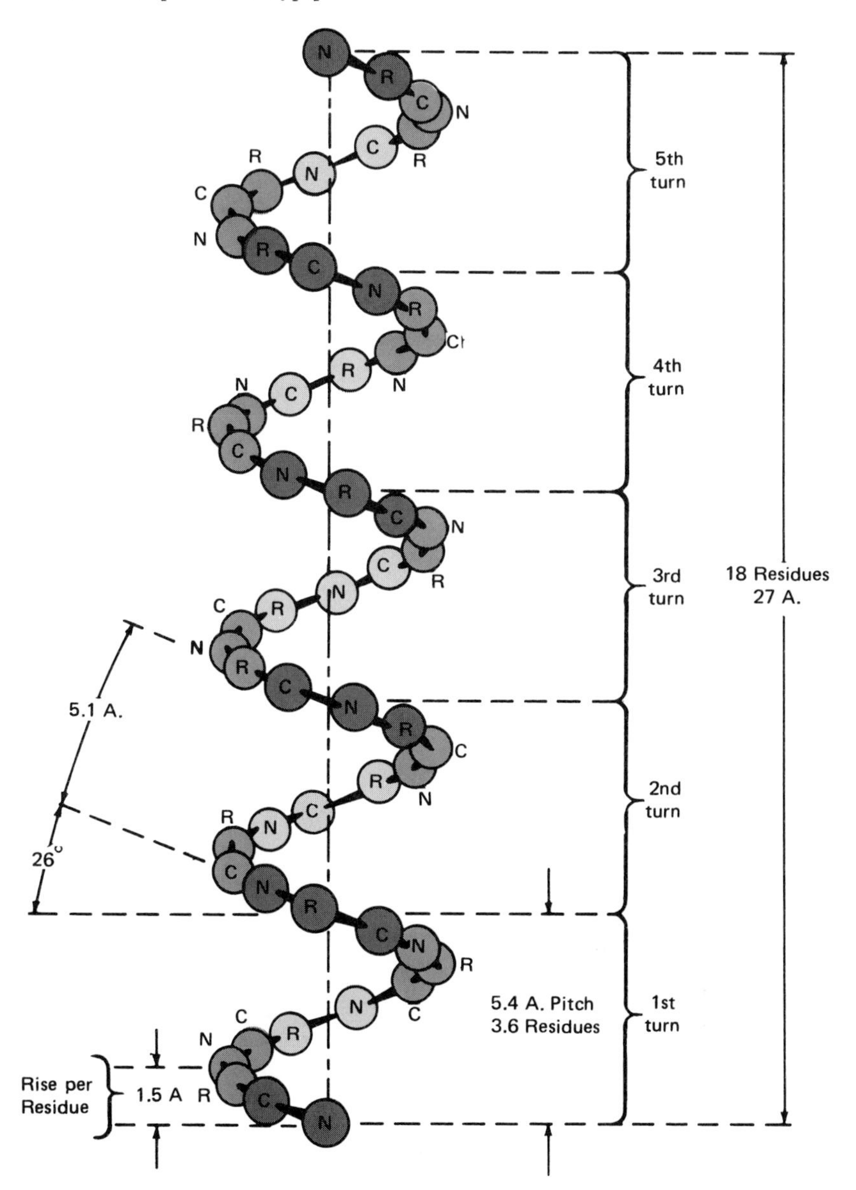

Figure 6.8. Detailed geometry of α-helical conformation. Reproduced with permission from C. B. Anfinsen *The Molecular Basis of Evolution,* John Wiley and Son.

The helical forms are very similar despite the great difference in the atomic constitution of the two kinds of molecules.

Another important ordered structure generated by polypeptides is one in which the chains are in a much more extended conformation. This is called the β-conformation, and the chain is almost in planar zigzag form. The structure of a single chain in this conformation is shown in the left-hand portion of Figure 6.10. Such

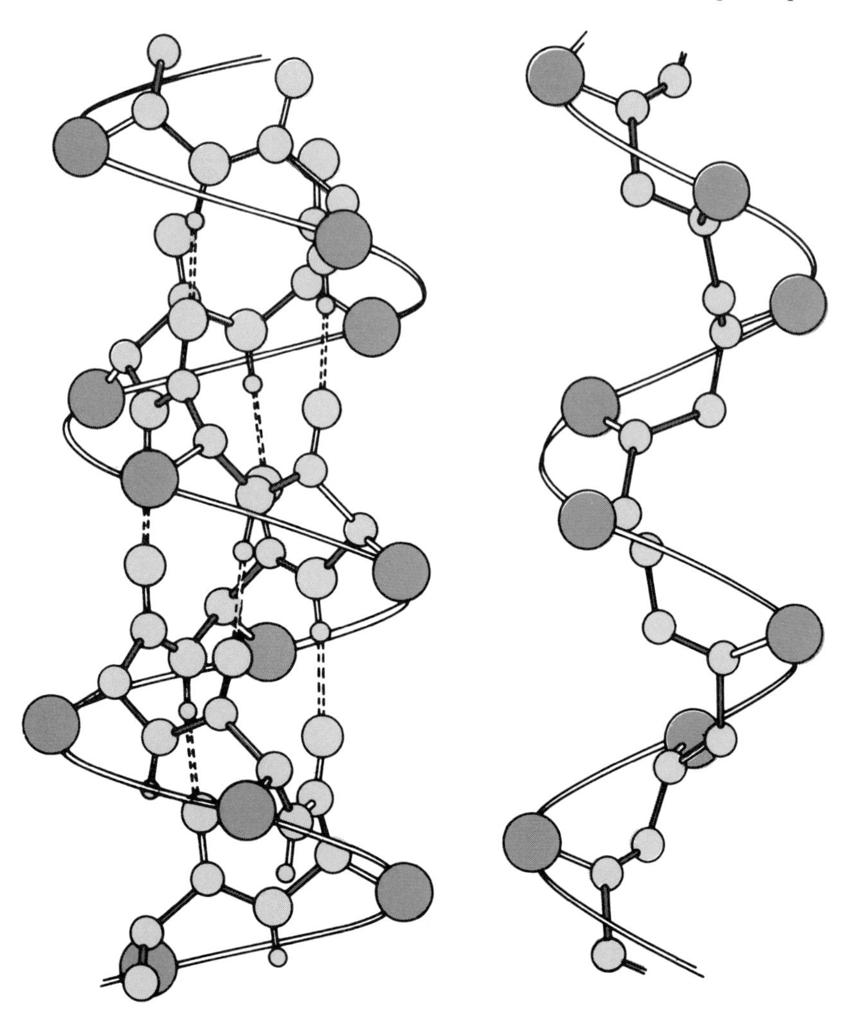

Figure 6.9. Comparison of the α-helix and helical structure generated by an isotactic vinyl chain. From G. Natta and P. Corradini, Rubber Chemistry and Technology *33*, 703 (1960). Reproduced with permission.

chains are very often organized in sheet-like structures, as is indicated on the right side of this diagram. Alternate chains can be placed either parallel to one another or can be oppositely directed or antiparallel. The distance between the chains is such that in order to achieve the most effective intermolecular hydrogen bonds (indicated by the dashed line in the figure), the individual chains are slightly pleated rather than being fully extended. The β-structure of the polypeptides is very similar to the ordered structure of the synthetic polyamides.

The third ordered structure is unique to poly-L-proline and results from the special chemical constitution of this polymer. This is also a helical structure, but it has a left-handed sense as compared with the right-handedness of the α-helix. A schematic diagram of this structure is given in Figure 6.11. This is an integral helix

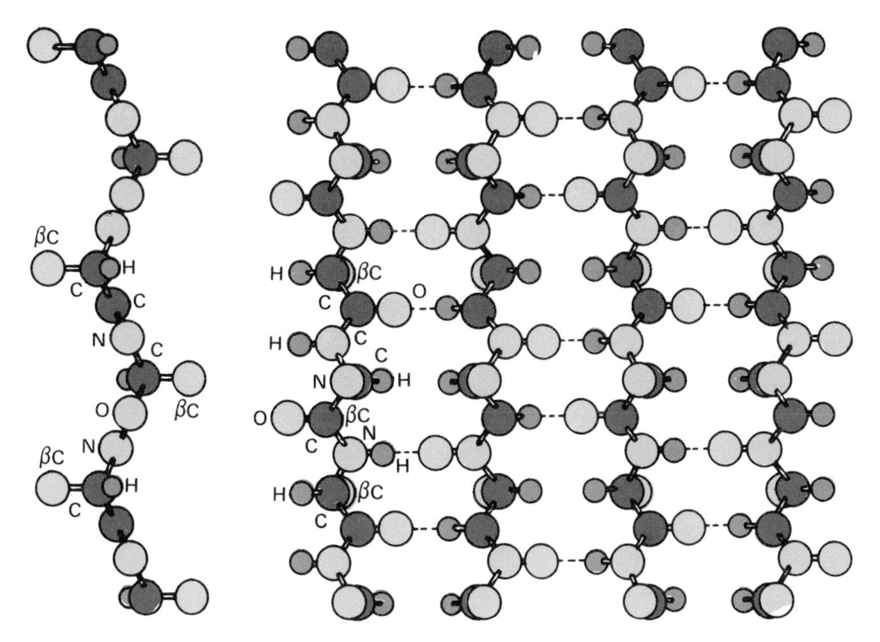

Figure 6.10. The β conformation of polypeptide chains. From R. E. Marsh, R. B. Corey and L. Pauling, Biochemica et Biophysica Acta *16*, 13 (1955). Reproduced with permission.

with three residues per turn. In contrast to the α-helix, the C=O groups are now directed almost perpendicular to the helical axis. Because of this orientation, these groups cannot interact with others along the chain. The stability of this helical structure arises solely from steric constraints.

In order to generate ordered polypeptide structures, it is not mandatory that the chain have chemically identical repeating units. Since the chain backbone is the same for all the amino acids, units with different side groups can be accommodated within the same structure. There are, however, some exceptions to this co-ordering or co-crystallization. For example, side groups having the same charge, and in close proximity, will repel each other and destroy the ordered structure. For steric reasons proline can only occupy the initial position of a right-handed α-helix and cannot occur within the body of such a helix. Cystine, with its disulfide cross-link, is also very difficult to accommodate in any of the ordered structures. A similar restriction holds for other residues with very bulky side groups. In copolypeptide chains, ordered structural sequences can be interspersed with disordered ones, depending on the composition and distribution of the amino acid residues.

Proteins–General

A protein is a polypeptide chain, or collection of chains, that has a very definite sequence of amino acid residues. This specificity of repeating units, with different side groups, gives the molecule its structural form, which in turn determines its

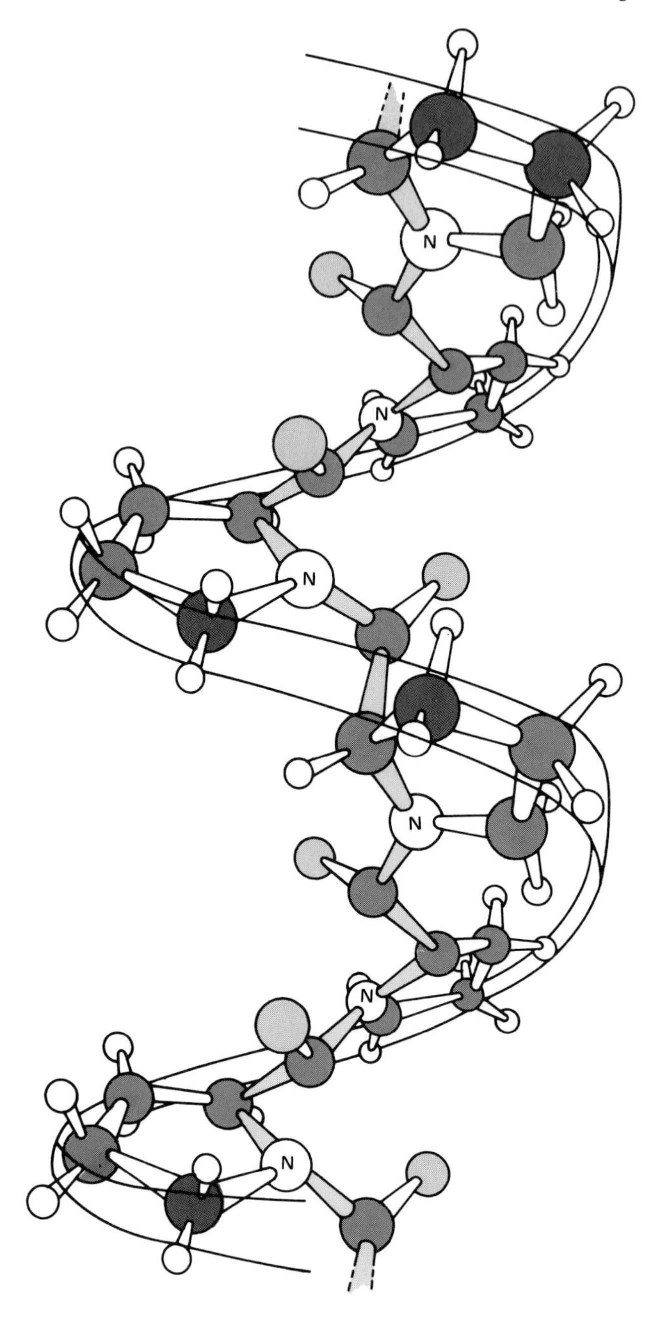

Figure 6.11. Representation of the ordered conformation of poly-L-proline.

function. Since 20 different amino acids are available as co-units, an astronomical number of chemically distinct molecules can be generated with a long-chain. Therefore, the same fundamental type of chemical repeating unit can participate in molecules that vary widely in their size, shape, and mass and also display a diversity of function. The range, scope, and vitality of proteins in man and lower organisms are truly remarkable. The specificity of a given protein is a result of evolutionary processes that took place over many billions of years. This selection and rejection process, by means of natural polymerization, leads eventually to the required sequence of amino acids. Proteins provide us with protection in the form of skin, hair, nails, and claws. Our connective tissue and tendons are proteins, as is the contractile apparatus of muscle and our veins and arteries.

Besides providing us with protection, physical structure, and mechanical capability, proteins are involved in a large number of chemical processes that are vital to our existence and well-being. In the form of enzymes they serve as catalysts for the biochemical reactions necessary to sustain life. They serve as carriers of oxygen and other small molecules and ions to various parts of the body. The globulins and antibodies with which we fight disease are also proteins. Cell membranes and our mitotic apparatus are also made up of this class of macromolecules. We might then properly ask how this all comes about. The answer to this question lies in the fundamental principles of macromolecular properties which has been our major theme.

A given protein can be characterized by its overall composition, i.e., the relative proportion of the different amino acid residues that are present. More important, however, is the sequential arrangement of these units along the chain. Starting from one end of the chain, conventionally the amino end, (NH_2), one can specify in principle, and also in practice in many cases, each successive amino acid residue until the terminal carboxyl (COOH) group is reached. The overall composition and sequential arrangement are, of course, different for the different proteins. The rotational characteristics of the single bonds in the chain backbone, moderated by the spatial expanse of the side groups and their interactions, yield a unique, very specific three-dimensional structure. We shall soon discuss in detail several examples of three-dimensional protein structures. At present, it suffices to say that this well-defined structure establishes the function of the protein through shape specificity and the location of reactive side groups.

To start a more detailed examination of these principles, we list the amino acid compositions of some typical proteins in Table 6.2. Virtually all of the amino acid residues are represented in each of the proteins. The overall composition varies widely, however, from one protein to the other. In some proteins all the amino acids are more or less equally well represented. In others, however, there is a very definite predominance of certain kinds of residues. We shall have occasion to refer to Table 6.2, as well as to the compositional characteristics of other proteins, as we examine specific cases in more detail.

The exact sequence of amino acid residues has been determined for a large number of proteins. The task still remains to be accomplished for many more, particularly those of very high molecular weight. An example of the sequence analysis for the enzyme ribonuclease is given in Figure 6.12. This protein, which has a molecu-

Table 6.2. Amino Acid Composition of Selected Proteins*

	Human Insulin	Horse Hemoglobin	Egg Albumin	Ribonuclease	Silk Fibroin	Wool Keratin	Collagen
Molecular Weight	6,000	64,500	45,000	14,000	–	–	–
Amino Acids							
Glycine	4	48	19	3	581	87	363
Alanine	1	54	35	12	334	46	107
Valine	4	50	28	9	31	40	29
Leucine	6	75	32	2	7	40	29
Isoleucine	2	0	25	3	8	0	15
Phenylalanine	3	30	21	3	20	22	15
Tryptophane	0	5	3	0	0	9	0
Proline	1	22	14	4	6	83	131
Hydroxyproline	0	0	0	0	0	0	107
Serine	3	35	36	15	154	95	32
Threonine	3	24	16	10	13	54	19
Tyrosine	4	11	9	6	71	26	5
Aspartic Acid and Asparagine	3	51	32	15	21	54	47
Glutamic Acid and Glutamine	7	38	52	12	15	96	77
Lysine	1	38	20	10	5	19	31
Arginine	1	14	15	4	6	60	49
Histidine	2	36	7	4	2	7	5
Cysteine	0	4	5	0	0	0	0
Cystine	3	0	1	8	0	49	0
Methionine	0	4	16	4	0	7	5

*When the molecular weight is known, the composition is expressed as the number of amino acid residues per molecule. When the molecular weight is not known, the composition is expressed as the number of moles of amino acid residues per 100,000 g of protein.

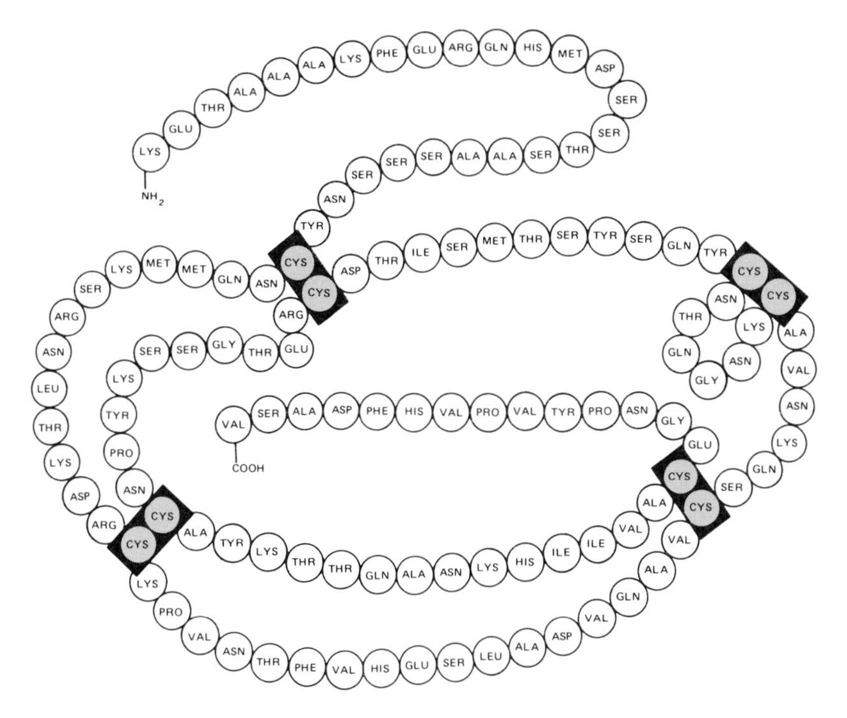

Figure 6.12. Sequence of amino acids in ribonuclease.

lar weight of 14,000, is composed of 124 residues arranged in the manner indicated. There are eight cystine groups in the single chain comprising the molecule. They are so positioned and the chain so organized that four pairs of disulfide (S–S) cross-links are made. In this case, the cross-links are intramolecular, i.e., within one chain or molecule. The diagram in Figure 6.12 is not intended to represent the real structure. It is just a convenient device by which to convey the sequential order of the repeating units. In a very notable achievement, ribonuclease has been completely synthesized in the laboratory, starting with the monomeric amino acids. The 124 amino acids were assembled in the required sequence. The resulting synthetic polymer possesses all the structural, chemical, and enzymatic properties of the native protein. The path now appears to be open for the synthesis of other proteins.

Fibrous Proteins

Proteins can be conveniently divided into two categories: globular and fibrous. The fibrous proteins follow the same structural and morphological definition that we have already given to the synthetic fibers. They are crystalline, and the chains are preferentially oriented in an axial direction. The ordered conformations found in the fibrous proteins correspond to the major ones we have discussed for the polypeptide chains. This does not mean that the complete protein molecule is crystalline and ordered. Among the more important fibrous proteins are the collagens, which are the proteins of connective tissue, of tendons, and of hide. The keratins, which give protective covering in the form of skin, fur, hair, feathers, nails, and scales are also fibrous proteins. Myosin is a fibrous protein directly involved in muscular contraction. Other fibrous proteins are epidermin, another skin component, and fibrin, the protein that forms blood clots. We shall focus our attention on three of the major fibrous protein systems: wool or hair, silk, and collagen.

The α keratins are a very familiar class of fibrous proteins. Two typical members are wool and hair. An examination of the amino acid composition of wool keratin, from Table 6.2, indicates the presence of most of the amino acids. There is, however, a very high concentration of cystine residues which is characteristic of the α-keratins and is indicative of a high level of intermolecular cross-linking. The primary structural unit of wool or hair is the α-helix. X-ray diffraction and electron microscope analysis make abudantly clear, however, that the ordered chains are not simply arranged side by side. Instead a very detailed organizational hierarchy exists.

The smallest organized entities that have been established are fibrous structures that lie along the complete length of the sample. These are about 20 Å in diameter and are called protofibrils. Each protofibril is probably composed of three α-helical chains twisted around each other so as to form a coiled-coil or supercoil. The chain structure within a protofibril is illustrated in Figure 6.13. Each strand in this figure

Figure 6.13. Schematic representation of three α-helical chains entwined around one another to form a protofibril of α-keratin. Adapted from L. Pauling and R. B. Corey, Nature *171*, 59 (1953).

represents an α-helix. The detailed organization of the protofibrils has been revealed by the electron microscope. As the cross-sectional view of Figure 6.14 shows, nine protofibrils are arranged in a circle about two others. Thus, an 11-stranded cable, approximately 80 Å in diameter is formed. This structure, or microfibril, is embedded in an amorphous, nonorganized protein matrix. Hundreds of such microfibrils are then joined together to form a fibrous bundle, or macrofibril, which is about 2000 Å in diameter. Thus, in the α-keratins there is an ordered progression of increasingly complex structures, starting with the α-helix of a single protein chain.

The physical properties of wool, hair, and the other α-keratins are derived from this morphological structure and from the peculiarities of the chemical composition. Within the protofibrils, the α-helices are intermolecularly cross-linked by disulfide bonds from the cystine residues. This represents a special feature of the α-keratins. The elastic properties of the keratins are qualitatively similar to the synthetic fibers and reflect the highly ordered chain structure. The extensibility is, however, tempered somewhat by the disulfide content. The low sulfur-containing keratins are more extensible than those containing large amounts. The latter are very inextensible.

The shrinkage of wool and hair when immersed in certain solvents is well known. Irreversible or irrecoverable shrinkage results from the permanent breaking of the intermolecular disulfide cross-linkages by these reagents. Reversible dimensional

Figure 6.14. Electron micrograph of cross-section of an α-keratin fiber. Insert is a schematic representation of the electron micrograph results. Reproduced with permission from B. K. Filshie and G. E. Rogers, J. Molecular Biology *3*, 784 (1961).

changes accompany the melting of the ordered structure and its recrystallization, while the intermolecular disulfide cross-links are maintained. This process follows the principles discussed in Chapter 5.

The permanent waving of hair is based on the dimensional and elastic properties of the cross-linked chains. In the first stage of the process the disulfide bonds are severed. The enhanced elasticity allows the hair to be shaped or curled into the desired form. The disulfide bonds are then reformed giving a "permanent" shape to the hair. The original process, developed many years ago, used curling irons whose high temperature caused the breaking of the cross-links. More sophisticated chemical methods have now been developed which are the basis for home permanents carried out at room temperature. One reagent severs the cross-links by chemical reaction with the disulfide bonds. The S–S bond is converted to two –SH species, separating the chains and allowing the hair to be more easily deformed. After the hair is set in the desired manner, this reagent is washed out. The addition of another reactant reforms some of the disulfide cross-links and gives a permanency to the structure.

In contrast to wool and the other α-keratins, there are no sulfur-containing amino acid residues in silk, which is composed mainly of Gly, Ala, and Ser residues with relatively small amounts of the more bulky side chains. All silks are not, however, chemically identical. The different species of the silkworm produce a fibrous protein that contains different proportions of amino acids with bulky side groups. These compositional differences in turn influence the physical properties of the fibers. In the most common form of silk, from the silkworm, *Bombyx mori*, there is a predominant six residue sequence, Gly-Ser-Gly-Ala-Gly-Ala, which repeats itself for long distances along the chain. This sequence accounts for a large proportion of the amino acid residues that are present.

The ordered portion of the polypeptide chains in silk are in the extended β-conformation illustrated in Figure 6.10. Neighboring chains are antiparallel to one another, and a pleated sheet is formed. The Gly residues lie on one side of the sheet and the Ser and Ala on the other. Although the ordered structure of silk is quite well defined, it is not the only determinant of properties. The amino acids with bulky side groups cannot be accommodated within this three-dimensional, ordered structure. Therefore, disordered amorphous regions coexist with the crystalline ones. The constitution of silk is analogous to that of a block copolymer. Different classes of amino acids reside in each block. One type can participate in the ordered crystalline array while the other cannot.

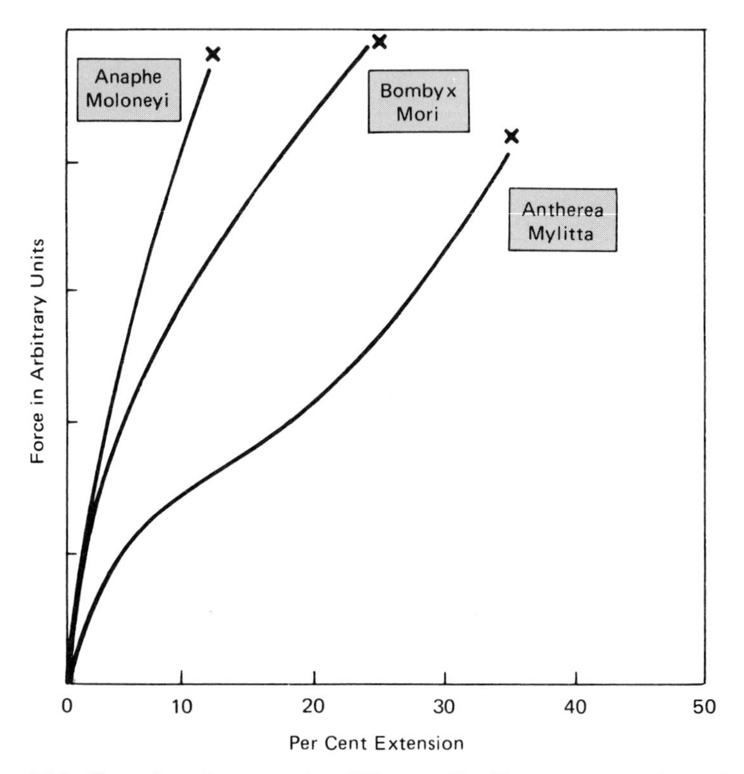

Figure 6.15. Force-length curves for different silk fibers. Adapted from F. Lucas, J. T. B. Shaw and S. G. Smith, Shirley Institute Memoirs *28*, 77 (1955).

The elastic properties of silk correlate very nicely with the fraction of bulky side groups present, and thus with the crystalline-amorphous ratio. Force-length curves for three different types of silk fibers are shown in Figure 6.15. Fibers from *Bombyx mori,* in which 13 percent of the residues have bulky side groups, are about 60% crystalline and can be extended 24% at its breaking point. On the other hand, in *Anaphe moloneyi,* only 5% of the residues are those with bulky side groups. It is, therefore, a highly crystalline, very rigid fiber that can be extended only 12.5% at its breaking point. The residues of *Antherea mylitta* contain 29% bulky side chains. It is the least crystalline of the silks and consequently can sustain the largest deformation. The same principles thus govern the physical properties of both the natural and synthetic fibers.

Collagen is one of the most abundant proteins in nature. In man it accounts for about one-third of the body protein. It is the principal fibrous component of cartilage, ligaments, and tendons and is also found in hide and in the eye cornea. It serves as a matrix upon which the mineral hydroxyapatite crystallizes to form bone. The outstanding property of collagen is its rigidity and resistance to deformation. Thus, nature has put it to use in bone, in strengthening animal skin and hide, and in connective ligaments and tendons where the mechanical force generated by muscle must be transmitted without loss. When collagen is dissolved in hot water, it is transformed into gelatin, which is much more soluble and is easily digestible.

We might expect that because of their special functions the collagens will have a unique amino acid composition. This turns out to be the case, as is revealed by the data in Table 6.2. On a gross composition basis all collagens contain about 33% glycine. Furthermore, every third residue is glycine, so that that amino acid is uniformly distributed. Another distinguishing compositional feature of this protein is the very high content of proline and hydroxyproline. The concentration of these residues varies with collagen type and species, and can be as high as 25 to 30% in some samples. Such high concentrations are unique among the proteins. Certain sequences of residues—such as Gly-X-Pro, Gly-X-Hypro—predominate in collagen. They form the basis for the particular ordered structure characteristic of collagen.

The fundamental structural unit of collagen is a very elongated molecule called tropocollagen. An electron micrograph of such molecules is shown in Figure 6.16. The tropocollagen molecule consists of three polypeptide chains intertwined with one another in such a way as to form the super helix shown in Figure 6.17. Each chain is essentially a polyproline left-handed helix (see Figure 6.11) with three residues per repeat. However, they develop a gentle twist to the right when they interact and combine to form the triple-stranded super helix of tropocollagen. The spatial arrangement of atoms in this particular array is such that two out of every three amide (NH) groups and carboxyl (CO) groups hydrogen bond to one of the other two chains as is indicated by the dashed lines in Figure 6.17. The distance between the chains and their periodicities are such that residues with side chains cannot be accommodated in every third repeating unit. This position must therefore be reserved for the Gly residue. Thus, the structure, composition, and distribution are consistent with one another. The bulky rings from the Pro and Hypro residues can fit into the structure without undue strain and without disrupting the hydrogen bonding pattern between chains. This particular ordered structure requires sequences

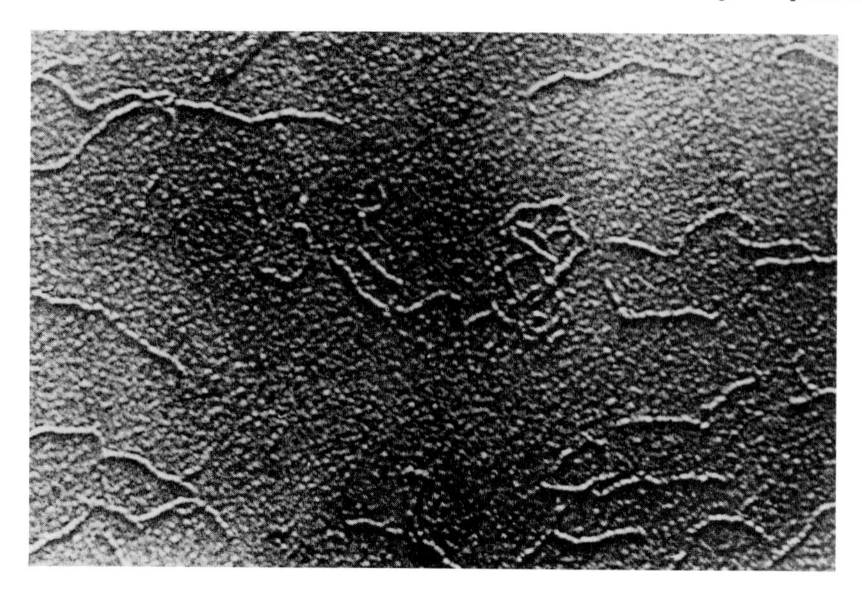

Figure 6.16. Electron micrograph of calf skin tropocollagen. Reproduced with permission from R. V. Rice, Proceedings National Academy of Science (U.S.) *46*, 1186 (1960).

of the type Gly-X-Pro, Gly-X-Hypro, and Gly-Pro-Hypro. Other sequences cannot be accommodated and would therefore not be found in ordered regions. The particular ordered structure of tropocollagen results from the unique amino acid composition and sequential distribution. In the molecular structure of tropocollagen one can already discern a source of the strength and rigidity of the collagen system.

The tropocollagen molecules form large macroscopic aggregates in different biological tissues. One kind of aggregate, found in skin and hide, is fibrillar as is shown in the electron micrograph of Figure 6.18. These fibrils are about 5000 Å in diameter and have a characteristic banded pattern that repeats every 640 Å. The banding results from the periodic absorption of the stain used in electron microscopy. This particular aggregate is formed by an overlapping set of perfectly alligned tropocollagen molecules, which are displaced one quarter of their length relative to one another.

The shrinkage of collagen fibers is a well known phenomenon that we have already discussed and illustrated in Chapter 5. There is a very strong correlation between the shrinkage temperatures in water for collagens from many different sources and the proline and hydroxyproline content. The greater the concentration of these residues, which contain the pyrollidine ring, the higher the shrinkage temperature. Upon further inquiry it is found that collagen molecules suitable to either cold or warm habitats are actually devised by nature through control of the amino acid composition. This concept becomes particularly striking when a comparison is made between cold-water and warm-water fish. Through control of the proline-hydroxyproline content, the collagen of warm-water fish have higher shrinkage temperatures than those whose natural habitat is the cold. The shrinkage

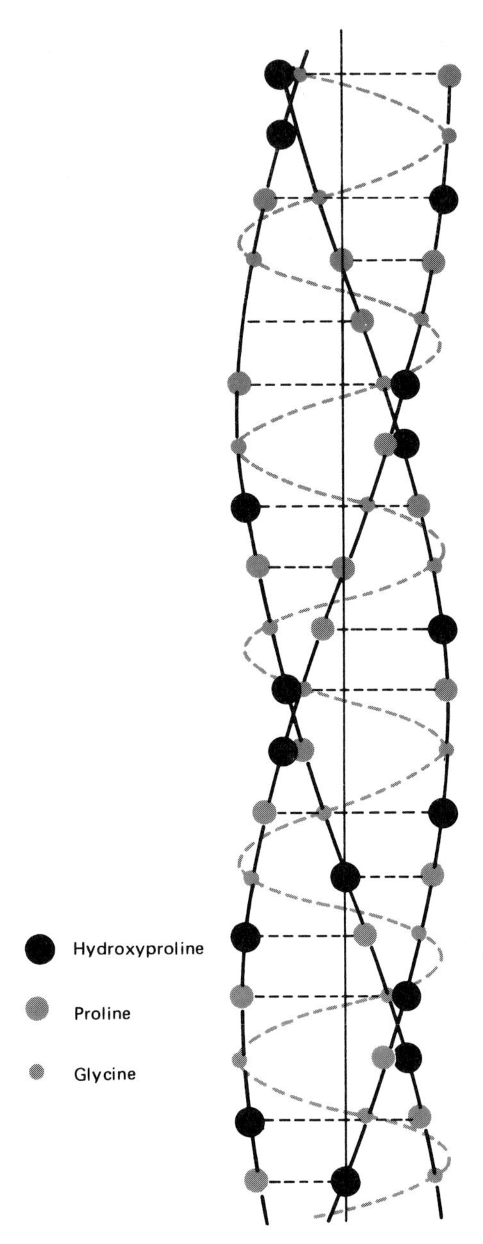

Figure 6.17. Schematic representation of the structure of tropocollagen. Three polypeptide chains are intertwined to give this structure. Reproduced with permission from J. R. Holum, *Principles of Physical, Organic and Biological Chemistry,* John Wiley and Sons, Inc.

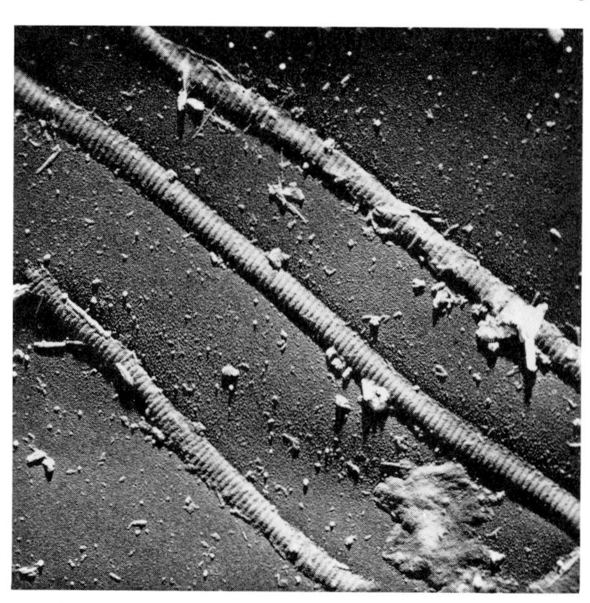

Figure 6.18. Electron micrograph of collagen fibrils. Reproduced with permission from R. V. Rice, Proceedings National Academy of Sciences (U.S.) *46*, 1186 (1960).

temperatures appear to be about 15° to 20° greater than the highest temperature likely to be encountered by a species.

Depending on the specific function to be performed, protein molecules are very often assembled into very intricate superstructures. These structures are much more complex than keratin and collagen fibers. An example of such a complex structure is the muscle system whose function in nature is to convert chemical energy into mechanical work. An electron micrograph of a muscle fiber from vertebrate skeletal muscle is given in Figure 6.19 (a). A schematic representation of this structure is given in Figure 6.19 (b). The unit betwen two Z lines is called a sarcomere, and it repeats itself periodically along the fiber axis. The major characteristics of a sarcomere are a set of thin filaments that are attached to membranes set 3 microns (30,000 Å) apart. The main constituent of the thin filaments is the globular protein actin, which is arranged in filamentous form. As is illustrated in the figure, the thin filaments are interdigitated by much thicker filaments in the central region between membrane pairs. These thick filaments are composed of the protein myosin, which is the principal protein of muscle. Myosin possesses all the structural characteristics of a fibrous protein. Following a nerve impulse, the sarcomere, or muscle, is activated by the release of calcium ions and in the ensuing reactions the sarcomere contracts, or shortens, as is indicated in the lower portion of the figure. The shortening process manifests itself by a further interdigitation of the thick and thin filaments without any outward signs of any strucutral changes. The molecular basis for this shortening process, from both a chemical and structural viewpoint, is under extensive investigation in many laboratories throughout the world. The problem is obviously a very complex one. However, research indicates that its resolu-

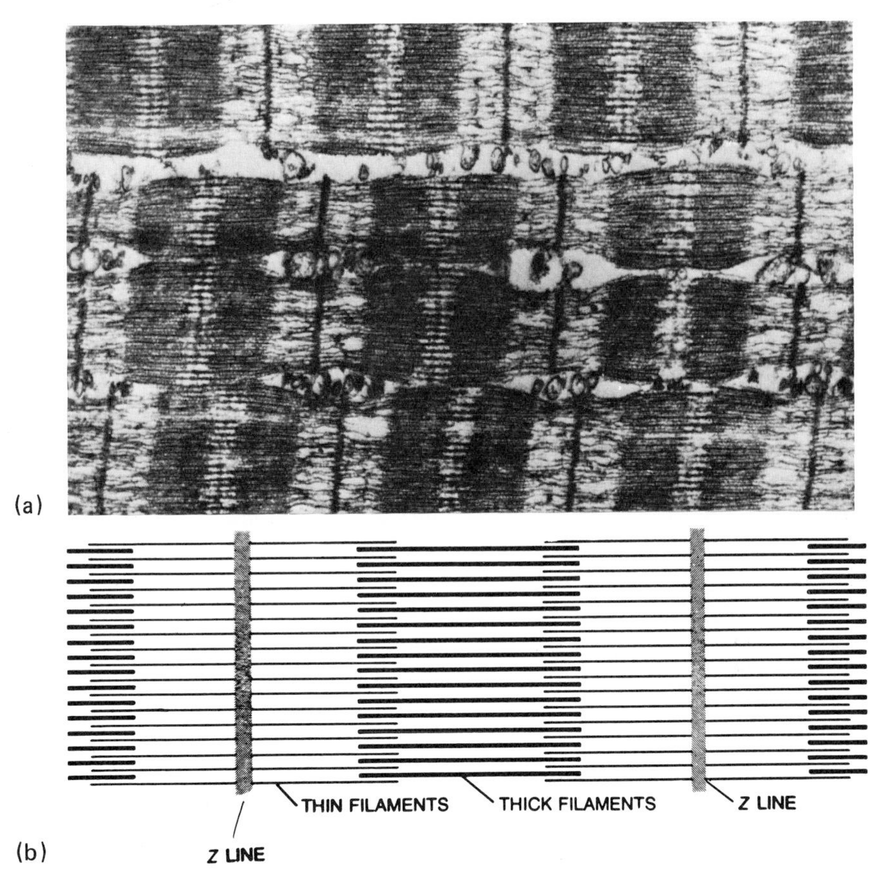

Figure 6.19. (a) Electron micrograph of muscle fiber from vertebrate skeletal muscle (top). Enlarged 25000 Å (b) Schematic diagram of muscle sarcomere (bottom). Reproduced with permission from J. M. Murray and A. Weber, *Scientific American 230 59* (1974) Copyright (1974) by Scientific American. Inc. All rights reserved.

tion will be found within the general principles of macromolecular science. Similar protein systems, assmebled perhaps in different ways, are responsible for motion and motility in other type cells and organisms.

Globular Proteins

It is evident from our discussion of the structure and properties of the fibrous proteins that they have very strong similarities with fibrous synthetic polymers. The globular proteins, however, present a unique molecular structure which we have not as yet encountered in our discussion of macromolecular systems. One of their distinguishing features is their compactness–virtually all the space within the domain on these globular molecules is filled and there is a negligible void volume. This structure is quite different from that of a chain molecule in a statistical conformation. The chain bonds do not adhere to a simple ordered conformation since the rotational angles do not repeat in a definite pattern. The structure of a given globular protein, although irregular, is also unique. Each globular protein represents a well-defined structure designed to carry out its specific function and mission. Polar, hydrophilic residues are found mainly on the exterior of the molecule, while the interior consists mostly of the nonpolar hydrophobic residues. The distribution of the residues is consistent with the fact that the natural habitat of the globular proteins is an aqueous medium.

The size, shape, and mass of the globular proteins vary over very wide extremes. Their molecular weights can be determined by standard methods. In some cases the molecules are so large that their outlines can be seen with the help of the electron microscope. The size and shape of the smaller molecules are deduced by methods of protein physical chemistry. Figure 6.20 gives a qualitative description of the shape and dimensions of some typical globular proteins, as well as their molecular weights. Many of the globular proteins can be crystallized from solution. Here the individual molecules are organized to form a three-dimensional, ordered array. When the molecules can be crystallized, then the methods of X-ray diffraction can be brought to bear on the problem to give a very precise (within 1 to 2 Å) location of all the atoms Thus, a very detailed structure of a very complicated molecule can be obtained.

In pursuing the premise that the sequence of amino acid residues determines the three-dimensional structure of a protein, we shall examine this relation for some of

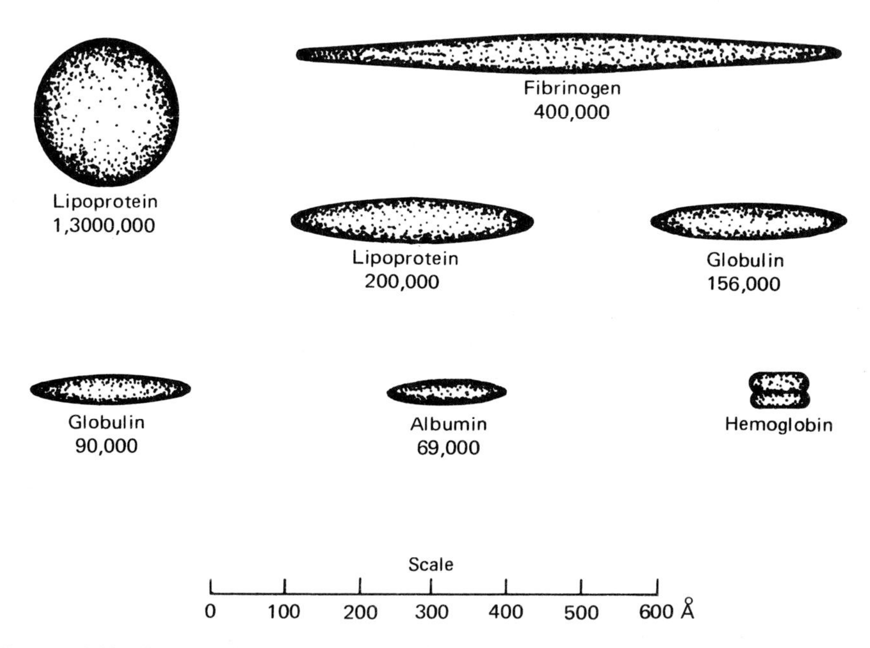

Figure 6.20. Shape, dimensions, and molecular weight of several globular proteins. Reproduced with permission from E. R. Harwick and C. M. Knobler, *Chemistry: Man and Matter,* Ginn and Company (1970).

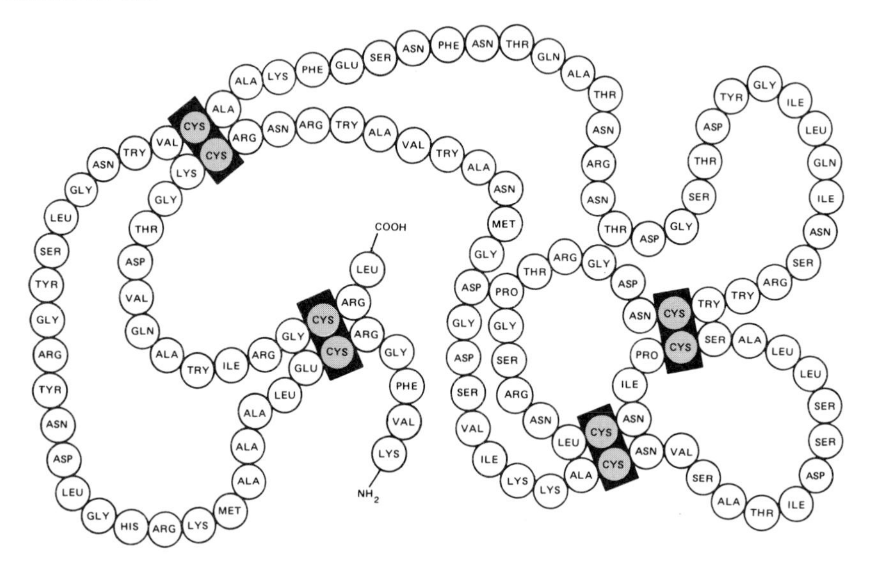

Figure 6.21. Sequence of amino acids in lysozyme.

the globular proteins who structure has been elucidated by X-ray diffraction. The structures of enzymes have now been detailed by this method. Enzymes are catalysts in that they increase the rate of a chemical reaction but are not consumed in the reaction. They are, therefore, very important in controlling chemical processes in nature.

One enzyme whose detailed structure has been elucidated is lysozyme, discovered by Alexander Fleming, who is also known for his discovery of penicillin. Lysozyme attacks many bacteria by dissolving, or lysing, their cell walls. This is accomplished by degrading a complex polymeric sugar or mucopolysaccharide. The enzyme catalyzes the breaking of specific chain bonds in the polysaccharide. Each molecule of lysosyme consists of a single polypeptide chain of 129 amino acid residues having a molecular weight of 14,600. There are four disulfide intermolecular cross-links in each chain. The sequential arrangement of the amino acid residues in the chain is shown in Figure 6.21. One, of course, cannot deduce very much about the real structure from this planar display.

A space-filling molecular model representing the three-dimensional structure is shown in Figure 6.22. The compactness of this molecule is readily apparent. Most of the polar, hydrophilic side chains are found on the surface of the molecule and can thus interact favorably with the surrounding water. The hydrophobic, nonpolar side chains are in the interior, for the most part, and are thus shielded from the solvent. The two major ordered polypeptide conformations, the α-helix and the β-sheets are found in some portions of the molecule. Other portions of the chain do not, however, have any recognizable ordered sequence, and the molecule as a whole has its own very special structure. The consequence of the sequential arrangement is the structure represented in Figure 6.22. The molecule is egg-shaped, and its most striking feature is the deep crevice running across its center. The whole strategy of stringing together this particular sequence of amino acids was to form this crevice,

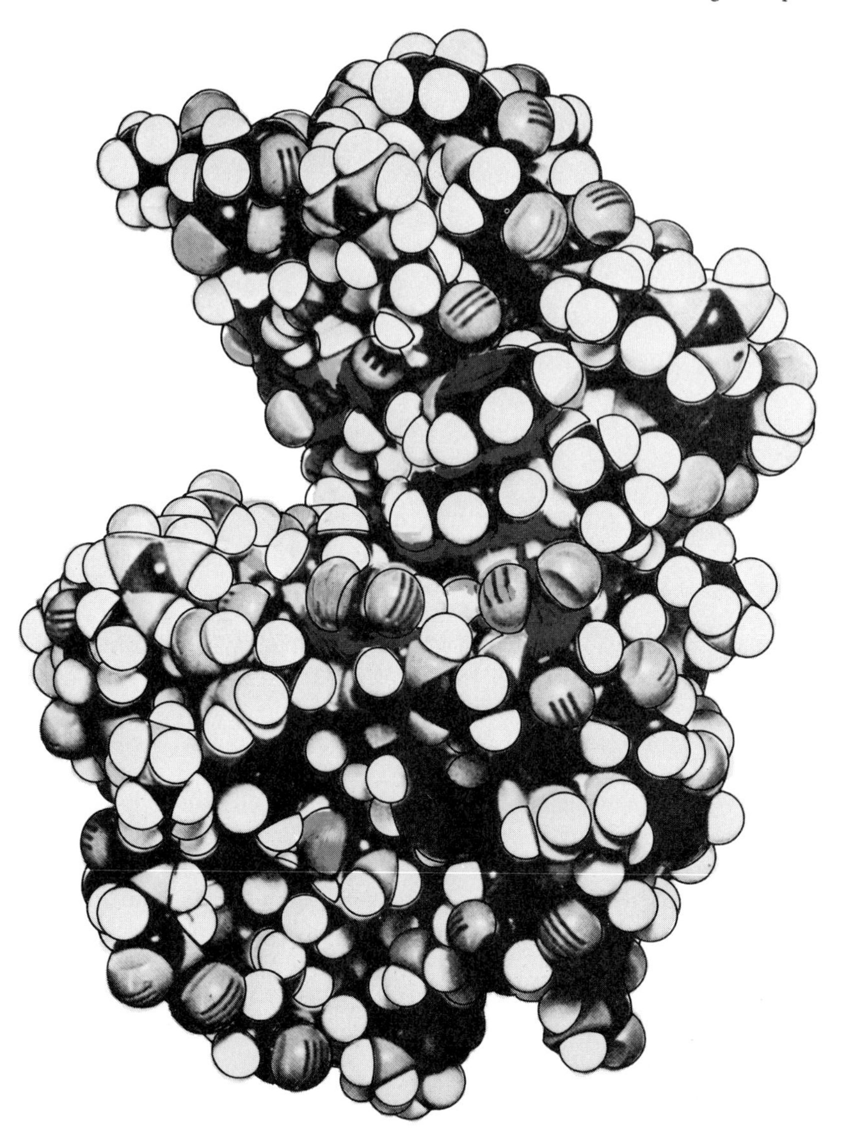

Figure 6.22. Molecular model of the three-dimensional structure of lysozyme. Reproduced with permission from R. E. Dickerson and I. Geis, *The Structure and Action of Proteins,* Harper and Row, Inc. Illustrations copyrighted by R. E. Dickerson and I. Geis.

for it is here that the enzymatic action takes place. The crevice, or cleft, on the enzyme surface is so designed that a portion of the sugar molecule (the substrate) fits very snugly into it. There is a very good geometric fit between the reactant and a portion of the enzyme surface. This fit is demonstrated in Figure 6.23 where the substrate has been placed into the catalytic trough. The amino acid residues that

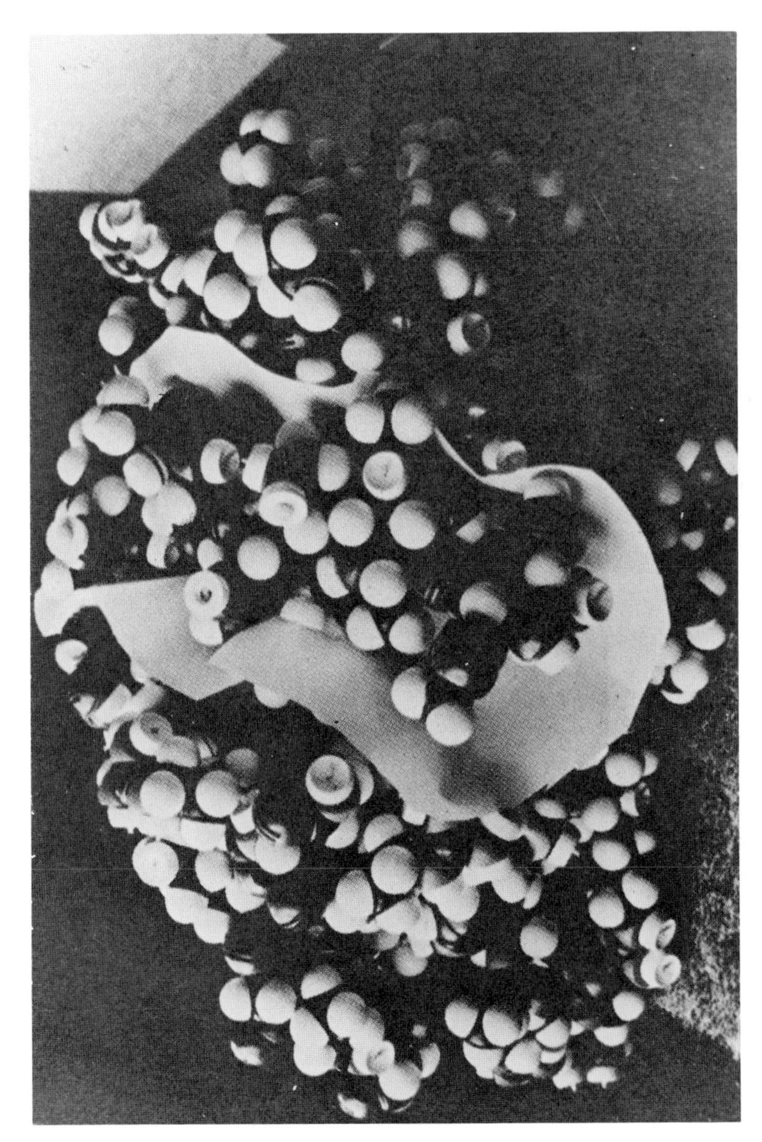

Figure 6.23. Three-dimensional model showing fit of the substrate (sugar) into the catalytic crevice of lysozyme. From *Concepts in Biochemistry* by F. J. Reithel. Copyright 1967. Used with permission of McGraw-Hill Book Company.

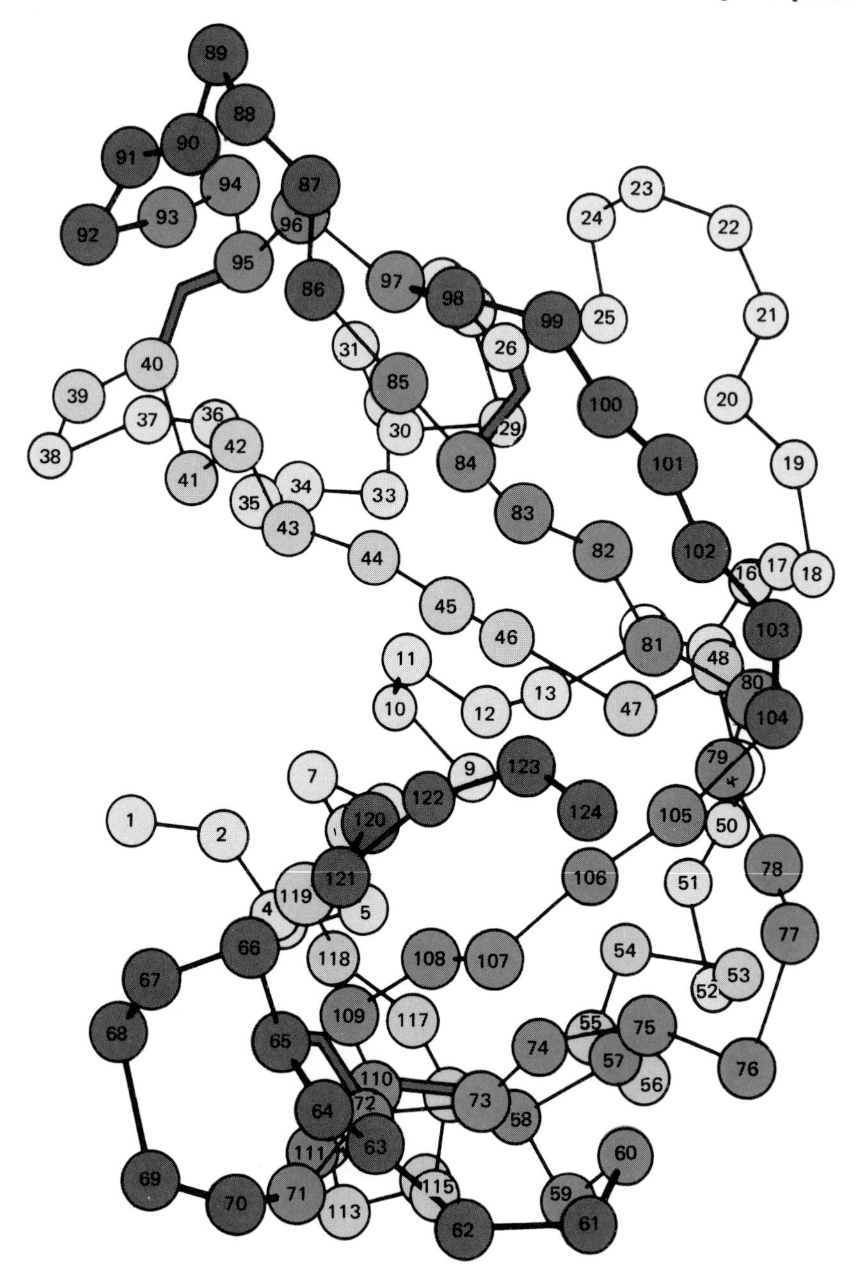

Figure 6.24. Representation of three-dimensional structure of ribonuclease. Reproduced with permission from R. E. Dickerson and I. Geis, *The Structure and Action of Proteins,* Harper and Row, Inc. Illustrations copyrighted by R. E. Dickerson and I. Geis.

participate in the reaction are found at the catalytic site. In this particular case residues 35 (glutamic acid) and 52 (aspartic acid) are involved. The design of this molecule, through the particular repeat of amino acids along the chain, has been to have it soluble in an aqueous medium, to present a proper geometric surface or catalytic trough for the particular sugar, and to have accessible the appropriate functional groups to carry out the reaction. For some classes of molecules it has been shown that motion occurs within the molecules so that the cleft is closed and the substrate surrounded. It is quite obvious that nature as a producer of polymers, designed for a specific chemical function, is far superior to the laboratory.

Ribonuclease is another enzyme that speeds up a chain-cutting reaction. However, instead of severing a polysaccharide chain, it performs the same function on a ribonucleic acid molecule (see next section). The amino acid sequence in ribonuclease has been given in Figure 6.12. The three-dimensional arrangement of these sequences is shown in Figure 6.24. In a general way the structure of this enzyme is similar to lysozyme. It is ovoid shaped with a deep crevice across its center. The crevice provides the proper geometrical fit for the nucleic acid. The functional groups necessary to carry out the scission are located at the surface of the crevice. These are residue numbers 12, 41, and 119. Although these residues are far apart in a linear array, they are brought relatively close to one another in the three-dimensional structure.

All the chain-cutting enzymes thus have certain major structural features in common. They are compact, oval-shaped molecules with deep crevices, depressions, on clefts so precisely designed that the reactant fits in and can react with the residues of the "active site." Many enzymes catalyze reactions that involve small molecules rather than polymer chains. Steric specificity will be the basis for the catalytic action here also.

In another set of functions the globular proteins act as transporters and storehouses for low molecular weight species. In vertebrates, one of the most important pairs of such molecules are hemoglobin and myoglobin. Hemoglobin, the main component of the red blood cells, binds molecular oxygen at the lungs and delivers it to myoglobin at the muscles. Here, the oxygen is stored until required for metabolic oxidative processes. On the return trip, the hemoglobin molecules help transport carbon dioxide back to the lungs. A single red blood cell contains about 280 million hemoglobin molecules, each having a molecular weight of 64,500. The oxygen-binding capability of this molecule is due to the heme group associated with it. This group, illustrated in Figure 6.25, consists of an iron atom surrounded by a porphyrin ring. The large number of double bonds in this ring results in a nearly planar heme molecule. The heme group also has characteristic colors. It is the oxygenated hemoglobin that is responsible for the deep red color of arterial blood; the bluish color of venous blood is due to the deoxygenated form of hemoglobin. In tissues, such as muscle, the red color is due to the oxygen attached to the myoglobin molecules.

Myoglobin consists of a single polypeptide chain of 153 amino acid residues having a molecular weight of 17,000. A single heme group is tightly bound to the chain, and the molecule can reversibly bind oxygen. The detailed structure of myoglobin

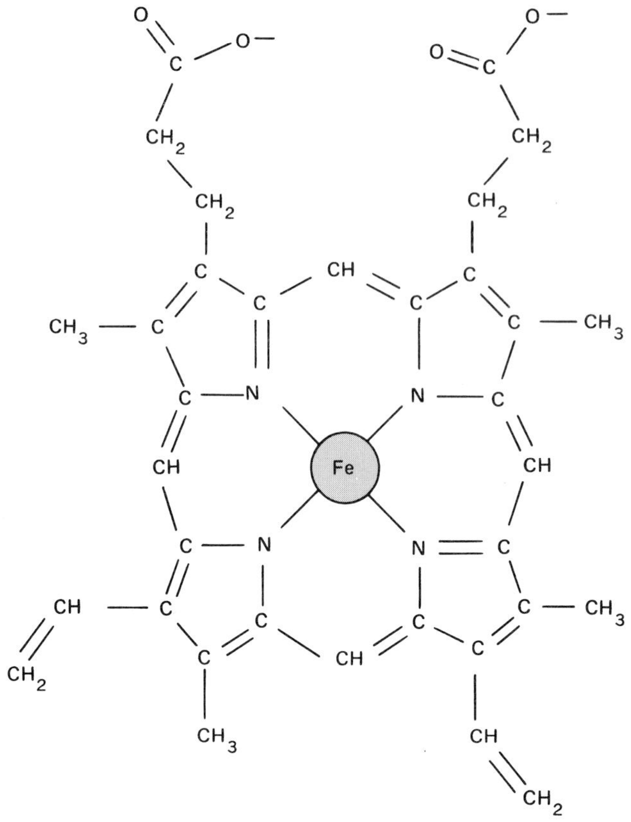

Figure 6.25. Skeletal structure of heme group.

has been elucidated by Kendrew and is shown in schematic form in Figure 6.26. The heme group is represented by the flat disc. The major function of the polypeptide chain is merely to serve as a container for this group. This protein has a very high proportion–about 80%–of its residues in the α-helical conformation. The container can be looked upon as being built up from eight connected pieces of α-helix.

Each hemoglobin molecule, the oxygen carrier, consists of two pairs of polypeptide chains. The α-chains each contain 141 amino acid residues, while the β-chains contain 146 residues each. The overall molecular weight of hemoglobin is 64,500. The four chains are so structured that they each entwine a heme group. The detailed structural analysis of hemoglobin has been worked out by Perutz and coworkers. Because of their pioneering work in determining detailed protein structures by the methods of X-ray crystallography, Perutz and Kendrew received the Nobel Prize for Chemistry in 1962.

The hemoglobin structure is shown in Figure 6.27. The oxygen-binding site is marked by O_2. Each chain has an irregular but well-defined structure very similar to that of myoglobin, and the chain pairs are symmetrically related. The packing of the four chains is such that there is close contact of side groups between unlike chains but virtually no contact between like chains. The overall molecular shape is

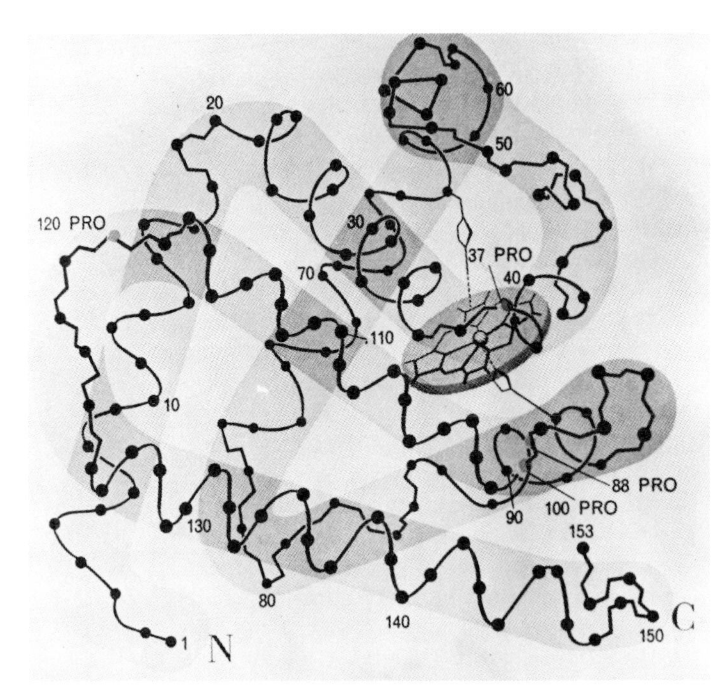

Figure 6.26. Schematic representation of three-dimensional structure of myoglobin. Reproduced with permission from *The Molecular Basis of Life,* W. H. Freeman and Co. Appeared originally in Scienfitic American.

roughly spherical. The hemoglobin molecule is more than just a receptacle for oxygen. It is so constructed that it can also control the amount of oxygen bound. All the details of the control mechanism have not as yet been worked out. However, it appears to involve a displacement of the β-chains. This movement alters the distance between the iron atoms, which in turn influences the amount of oxygen bound.

Outstanding advances have been made in our understanding of the molecular basis of some diseases. What is involved here is the manufacture or synthesis of molecules that differ from the normal ones found in a healthy person. The hemoglobin molecules show some very striking examples of molecular diseases. In some anemic patients there are abnormal α-chains or abnormal β-chains. The iron atom, which is usually in the ferrous state, in converted to the ferric state. When this occurs, the heme group loses its power to combine with molecular oxygen. This type of anemia is called ferrihemoglobinemia. Much more subtle differences in chain structure are also known to cause disease. For example, the sixth residue from the amino end in the β-chain of a normal hemoglobin is glutamic acid (Glu). Persons suffering from sickle cell anemia have valine (Val) substituted for Glu at this position. Glutamic acid has a carboxyl group, –COOH, on its side chain and thus can be ionized. On the other hand, valine has a hydrocarbon isopropyl side group, which cannot ionize. When the same glutamic acid is replaced by lysine, thalassemia results. When glutamic acid in the 26th position is replaced by lysine, a mild hemolytic anemia results. It seems almost incredible that a single amino

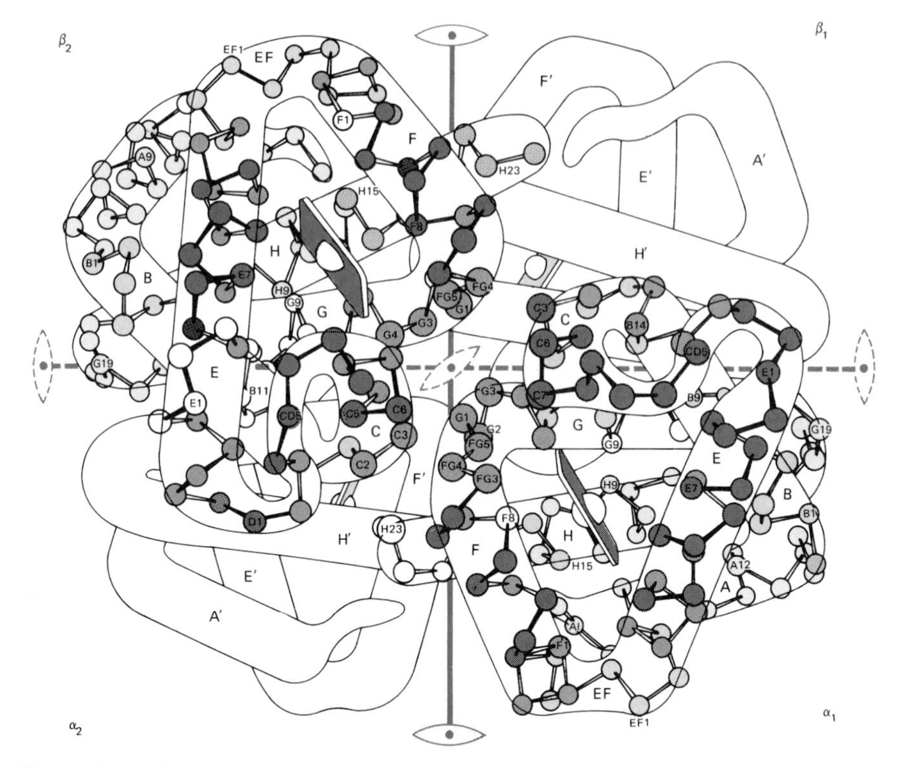

Figure 6.27. Schematic representation of three-dimensional structure of hemoglobin. Reproduced with permission from R. E. Dickerson and I. Geis, *The Structure and Action of Proteins,* Harper and Row, Inc. Illustrations copyrighted by R. E. Dickerson and I. Geis.

acid residue in the wrong place, which does not cause any major changes in the protein structure, could have such a devastating effect on normal biolgical function and result in serious disease. The study of diseases caused by the alteration of molecular structure is a relatively new endeavor. It has great promise in understanding and subsequently conquering many types of diseases.

The results cited above show that the biochemical machinery that synthesizes protein does not always function at its optimum. To learn more about this process, as well as about the transmission of genetic information, we have to study the properties of the nucleic acids.

Nucleic Acids–General

It has been known for almost a century that hereditary information is transmitted from one generation to another by the chromosomes of the cells. The region of the chromosome that determines a particular characteristic or trait is called a gene. It has also been established that protein synthesis is controlled by the genes. A gene therefore possesses a dual function central to life: it must be able to replicate itself so that genetic information is transmitted to successive generations and must be able to direct the synthesis of proteins so that the organism can function. The fundamental chemical constituent of the genes is a long-chain nucleic acid, deoxyribonucleic acid, popularly called DNA. Although DNA is the carrier of the genetic information and contains the message directing the protein synthesis, the actual assembling of the appropriate amino acids and their polymerization is carried out by another class of polymeric nucleic acids, the ribonucleic acids, or RNA.

The underlying molecular basis for reproduction and protein synthesis lies in the chemical composition and structure of the nucleic acids. These vital biological functions will be seen to be natural consequences of the chain-like character of these polymers and the chemical nature of their repeating units. A schematic diagram of the nucleic acid chain is shown in Figure 6.28. The backbone of the nucleic acid chain consists of repeating sugar units connected by phosphate bridges. Attached to each of the sugars is an organic base, either a purine or a pyrimidine. The major chemical distinction between the DNA and RNA chains is in the kind of sugar unit involved. Although in both cases the sugar is a five-membered ring it is D-ribose in RNA and deoxy-D-ribose in DNA as is shown in Figure 6.29. The deoxy-D-ribose sugar contains three hydroxyl groups as compared to the four that are present in the D-ribose molecule. For either sugar, when inserted in the appropriate nucleic acid, the hydroxyl group in the carbon-1 position is combined with one of the constituent heterocyclic bases. The chain is perpetuated by phosphate-hydroxyl linkages as is indicated in Figure 6.28 and involve the carbon atoms in the number 3 and 5 positions.

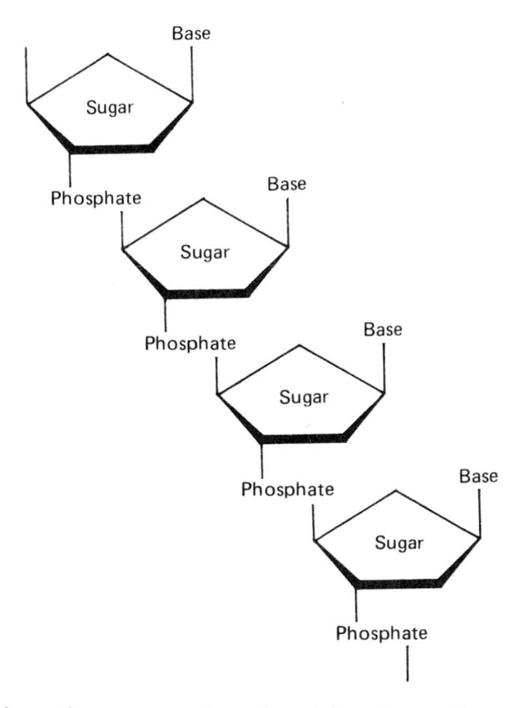

Figure 6.28. Schematic representation of nucleic acid chain.

D-Ribose Deoxy-D-Ribose

Figure 6.29. Structural formulas for D-ribose and deoxy-D-ribose.

The purine bases attached to the sugars are either adenine (A) or guanine (G). Their structures are represented in Figure 6.30. The pyrimidine bases are cytosine (C), uracil (U), and thymine (T). Their structures are shown in Figure 6.31. Cystosine and thymine are found in DNA, while in RNA uracil replaces thymine. The last

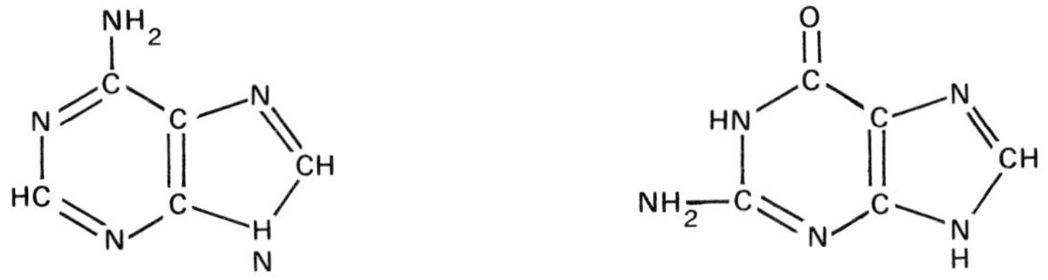

Adenine (A) Guanine (G)

Figure 6.30. Structural formulas of purine bases.

two bases differ by the presence or absence of a methyl $-CH_3$ group. In each of the nucleic acids, therefore, four chemically different bases are found. Figures 6.30 and 6.31 also show that the pyrimidine bases are much smaller than the purines.

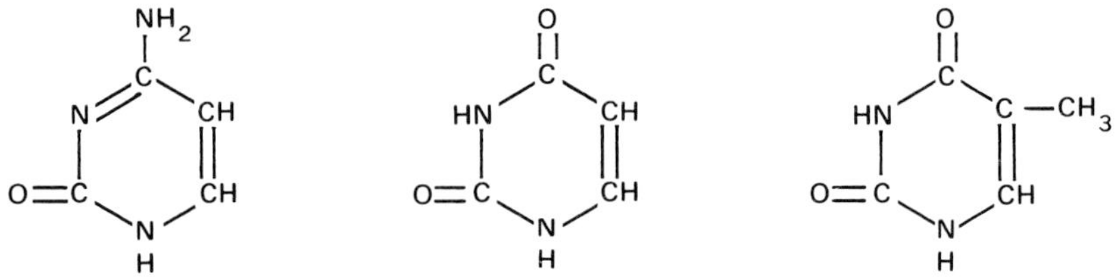

Cytosine (C) Uracil (U) Thymine (T)

Figure 6.31. Structural formulas of pyrimidine bases.

Detailed skeletal structures of both types of nucleic acid chains are given in Figure 6.32. The chain of nucleotides, as it is called, is perpetuated by a phosphate-ester linkage. One can imagine the polymerization to take place by the reaction between phosphoric acid and the hydroxyl groups of the sugar, which are associated with the carbon atoms in the number 3 and 5 positions. Thus, the sugar acts as

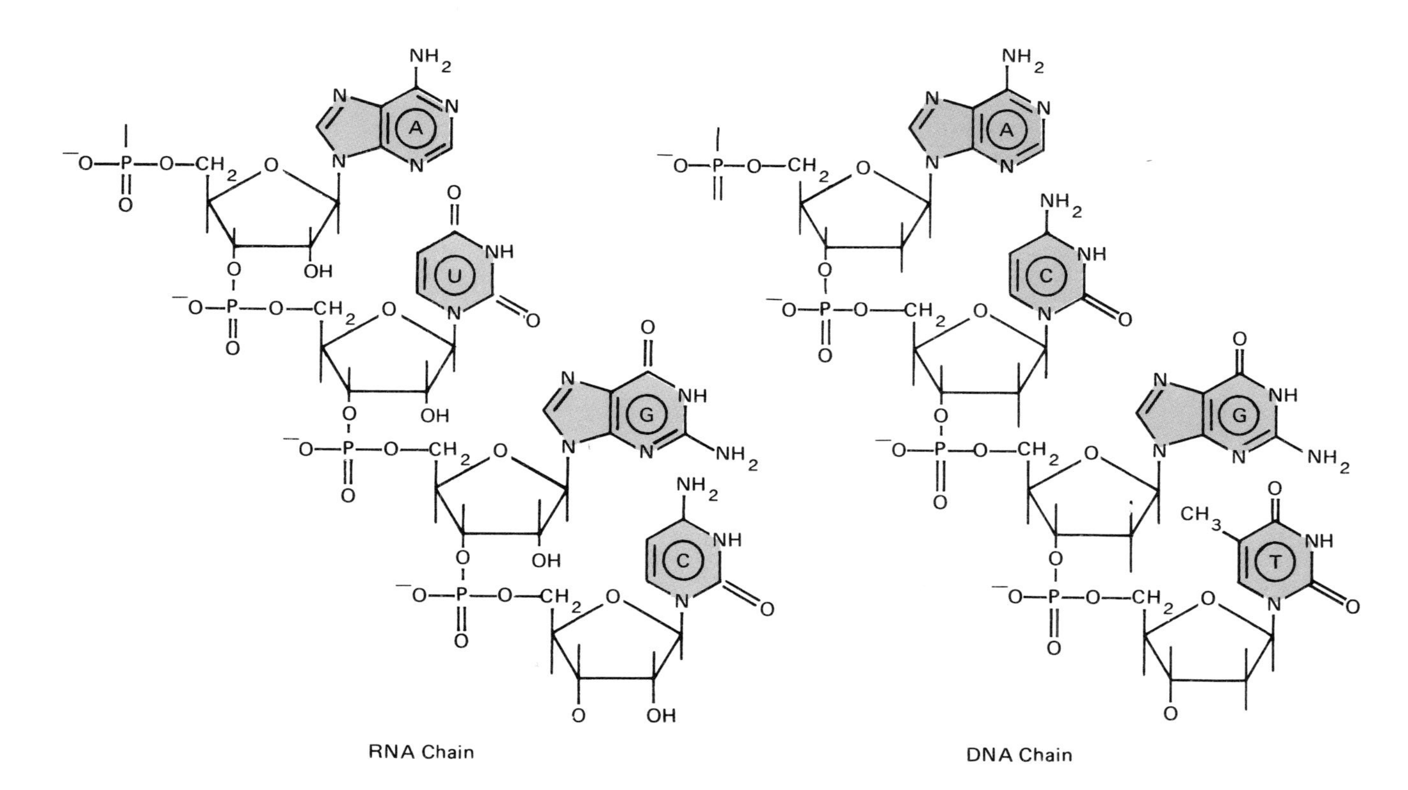

Figure 6.32. Skeletal structure of representative portions of RNA and DNA chains.

a bifunctional molecule in the polymerization process. In addition to the usual conformational versatility that can be expected in all chain molecules, there is strong chemical specificity inherent in the nucleic acids through the composition and sequential distribution of the bases.

Deoxyribonucleic Acid–DNA

The molecular weight of the genetic nucleic acid, DNA, is extremely high. Depending on the species, the molecular weight can range from several million to several hundred millions. These molecular weights correspond to several thousand to hundreds of thousands of nucleotide repeating units. The extended length of a DNA molecule from the chromosomes depends on the complexity of the organism from which it comes. For example, it has been estimated that if the DNA in a human cell were stretched out, it would make a thread about three feet long. It is, however, so compacted in the cell that dimensions are actually in the submicroscopic range.

To be able to store and to transmit genetic information, the chain must be capable of having a code built into its structure. We must, therefore, seek to elucidate both the code and the mechanism by which it is transmitted form one generation to another. Since the chain backbone is made up of identical repeating units, it is difficult to see how any information can be stored or encoded in this part of the structure. However, there are four different bases attached to the sugar. Their concentration and sequential distribution are in fact the only variables in the chain structure. The possibility thus exists that if the bases were organized in definite sequences, these arrangements could constitute the genetic code. The very high molecular weight of DNA also suggests the capacity to store a great library of information.

The first substantial clue that the bases are actually involved in the coding process came from a very comprehensive chemical analysis of DNA. This analysis established some very general relationships for the base composition. It was found by Chargaff, from a study of DNAs from many sources, that the number of adenine (A) bases always equals the number of thymine (T) bases. Furthermore, the number of guanine (G) bases is equal to the number of cytosine (C) bases. Consequently, the ratio of the total purine to the pyrimidine bases must be unity. On the other hand, the ratio (A+T):(G+C) in highly variable. Thus, there are two distinct, identifiable pairs of purine and pyrimidine bases. The amount of each of the bases making up a particular pair is the same. However, the relative amount of each of the pairs is variable. The universality of these relations for all DNAs makes it a quite significant observation and of underlying importance to DNA function.

The next problem that arises is how these compositional rules endow DNA with its special properties. The answer resides in the particular ordered structure of the DNA molecule. As we have seen from our discussion of synthetic polymers, polypeptides, and proteins, the details of ordered conformations can be obtained by the methods of X-ray diffraction. From the analysis of fiber patterns of highly oriented DNA, which were produced by M. H. F. Wilkins, Crick and Watson deduced their classical double-stranded helical model of DNA.[1] A space filling molecular model of the Watson-Crick DNA structure is given in Figure 6.33 (a) and in a schematic, and perhaps somewhat more illuminating form in Figure 6.33 (b). The Watson-Crick model of DNA consists of two complementary chains that run in opposite directions and are coiled around each other to form a double helix. It gives the appearance of a spiral staircase, or twisted ladder. The sides or banisters are made up of the phosphate-sugar chain backbone. The steps or rungs consist of two bases, one from each chain, which lie in a plane perpendicular to the chain axis. The distance between the chains is such that one of the bases must be a purine (A or G) and the other a pyrimidine (C or T). A pair of purine bases is too large to fit into the allocated space, while a pair of the smaller pyrimidine bases will leave a gap between the chains. We can see immediately that these requirements for the ordered structure exactly fultill the demands of the base competition, namely that the number of A bases must equal the number of T while the number of C bases must equal the number of G. Thus each base pair, which forms the rungs in the ladder, is made up of either an A–T or C–G combination. This condition simultaneously satisfies both the steric and chemical requirements but still allows a great deal of freedom in the base sequences since there are no restraints on the ratio (A+T):(G+C). The similarity in the dimensions of these two allowable base pairs is shown in the diagram of Figure 6.34. Besides satisfying the dimensional requirements, additional stability is given to the ordered DNA structure by these particular pairs by a set of very favorable hydrogen bond interactions, which are indicated by the dashed lines.

When this ordered structure is examined in more detail, we find that the two polynucleotide chains are twisted around each other in such a way as to form a right-handed helix that completes one turn every 34 Å. The base pairs, looking like a set of stacked pennies, are spaced 3.4 Å apart so that there are ten base pairs per turn of the helix. The diameter of the molecule is such as to be able to accommodate only the two particular sets of base pairing that we have described. The double-stranded molecule has a molecular weight of 196 /Å unit of length, or approximately two million per micron.[2] In higher organisms DNA molecules as long as 1000 to 2000 microns have been observed and correspond to fantastically high molecular weights. An electron micrograph of an elongated DNA molecule is shown in Figure 6.35. We have already noted that if the DNA molecule were completely stretched out in the human cell it would be extremely long and would have dimensions well above the cellular level. The DNA molecule, therefore, must be capable of existing in some type of coiled or compacted configuration while the detailed double-stranded molecular structure is maintained. Figure 6.36 is a representative electron micrograph of an isolated, ordered, double-stranded DNA molecule. We see that it is indeed capable of existing in a crumpled, compacted structure while its chemical and local ordered structure, the Watson-Crick double-stranded helix, is maintained.

[1] In their original paper, in 1953, Crick and Watson suggested that this structure had certain novel features of considerable biological interest. They in fact opened a major pathway of molecular biology, a scientific discipline that seeks to explain biological function in terms of molecular structure.

[2] One micron corresponds to 10,000 Angstroms or 10^{-4} cm.

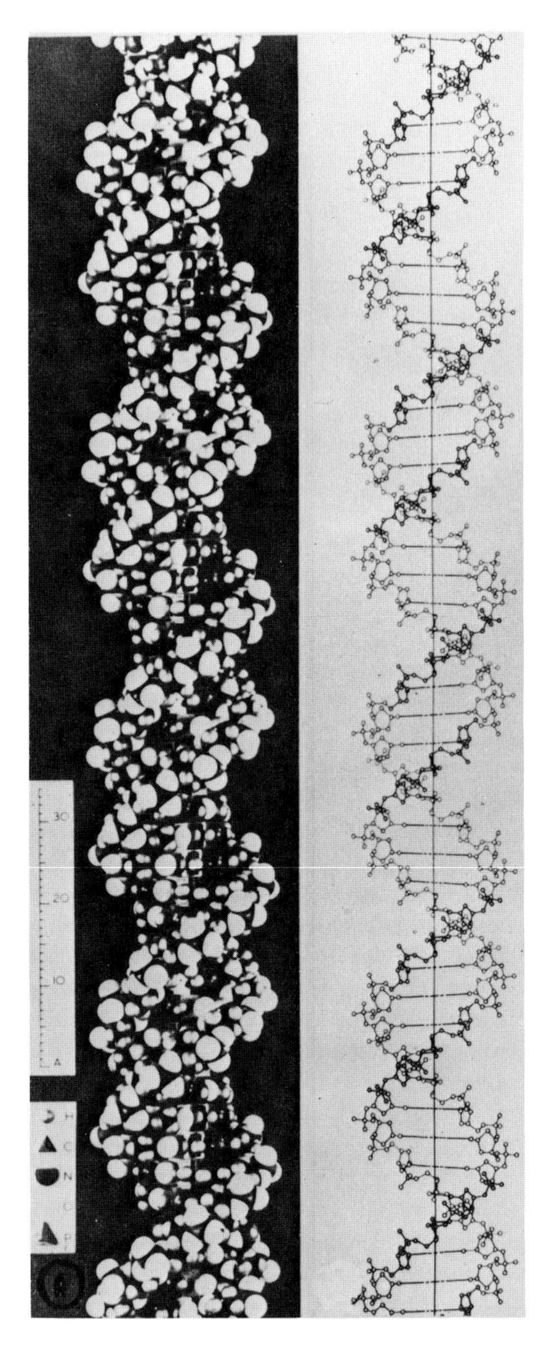

Figure 6.33. (a) Molecular model of ordered DNA structure. (b) Schematic representation of ordered DNA structure. Reproduced with permission from E. J. DuPraw, *Cell and Molecular Biology* (1968) Academic Press, N. Y.

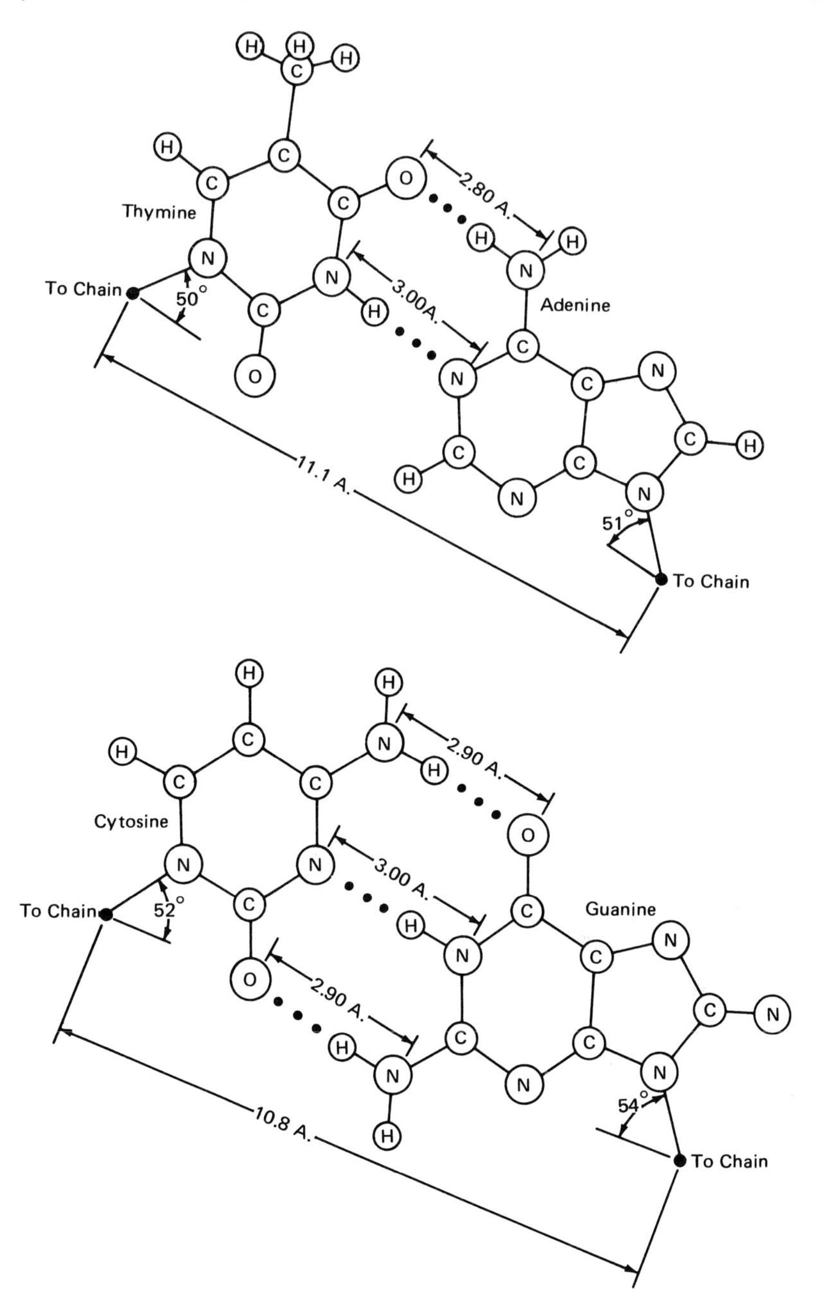

Figure 6.34. Dimensions and hydrogen bonding capabilities of base pairs. Reproduced with permission from C. B. Anfinsen, *The Molecular Basis of Evolution,* John Wiley and Sons.

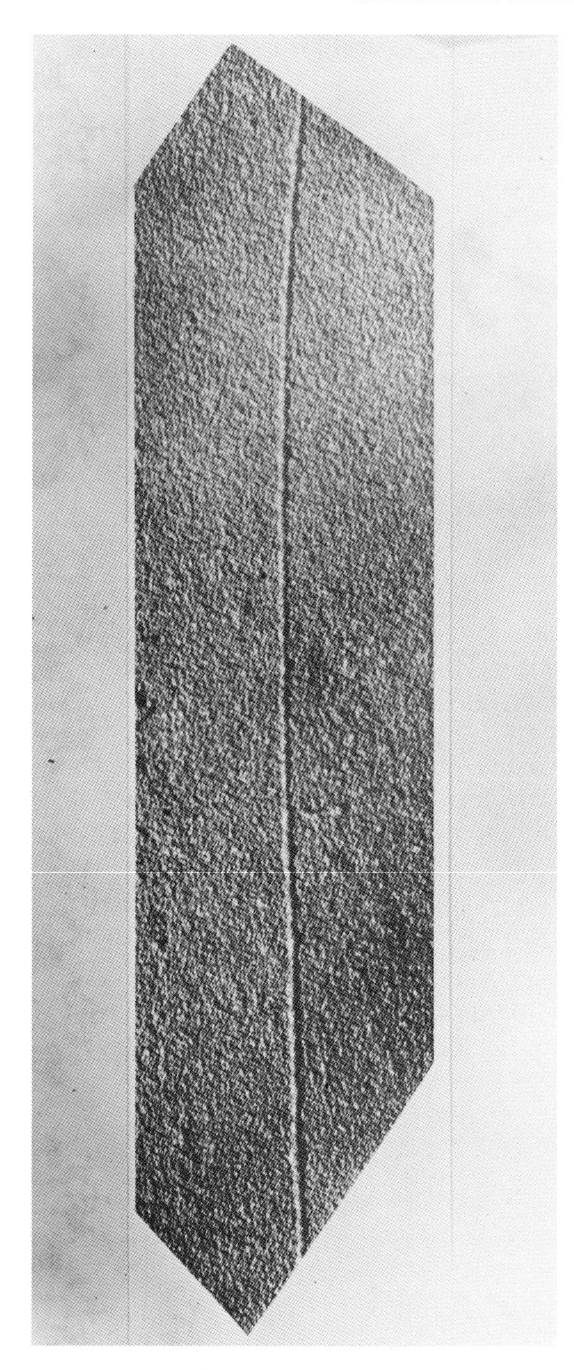

Figure 6.35. Electron micrograph of elongated DNA. Reproduced with permission from E. J. Dupraw, *Cell and Molecular Biology* (1968) Academic Press, N.Y.

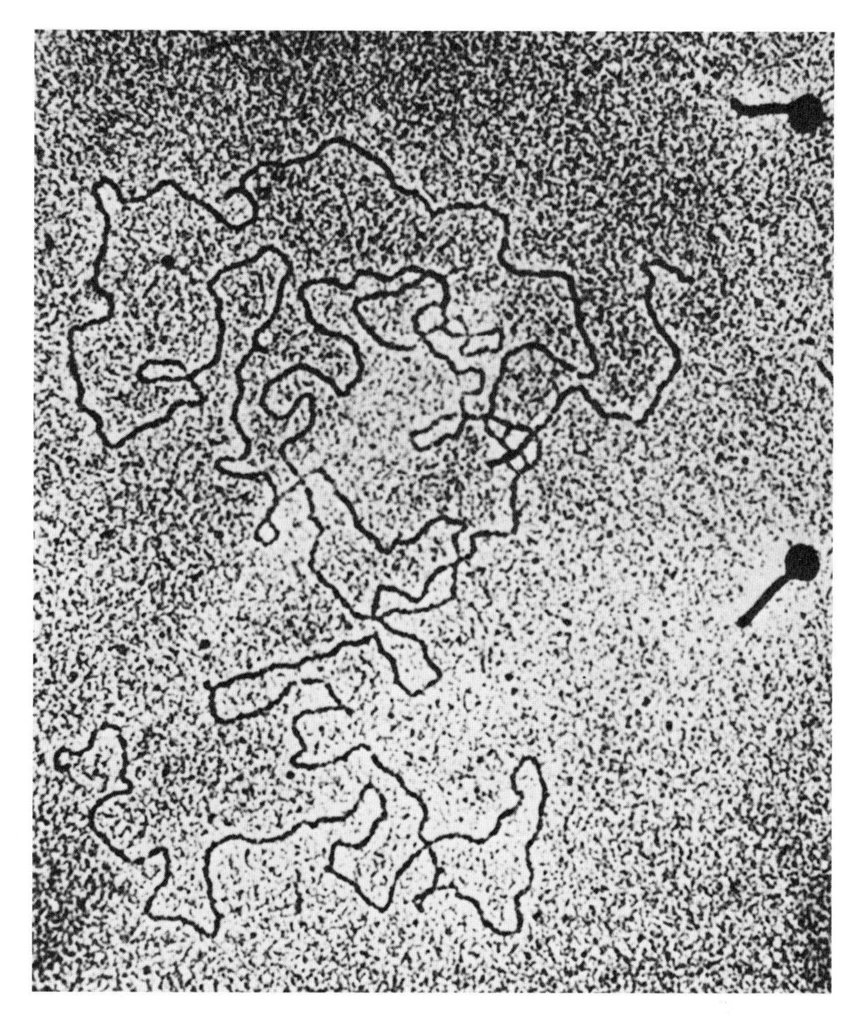

Figure 6.36. Electron micrograph of native DNA in compacted form. Reproduced with permission from C. Grobstein. *Scientific American* July 1977 pg. 27. Copyright (1977) by Scientific American Inc. All rights reserved.

The chemical and structural requirements for specific base-pairing are the key to the genetic function of DNA. Suppose, for example, that the four different bases were distributed along a single DNA chain in the manner indicated in the left-hand side of Figure 6.37. However, since only particular bases can be linked with one another, the sequence of bases on the other chain in the molecule is immediately fixed by the sequence of the first chain. Thue, in Figure 6.37, A must be linked to T, T to A, G to C, T to A, and so on down the line. One chain thus serves as a template for the base sequence of the other. Since the two chains are complementary, rather than identical, they can have widely different base compositions.

We can now visualize biological replication at a molecular level in a very definite way. When a cell is replicated it is necessary to provide the new daughter cell with

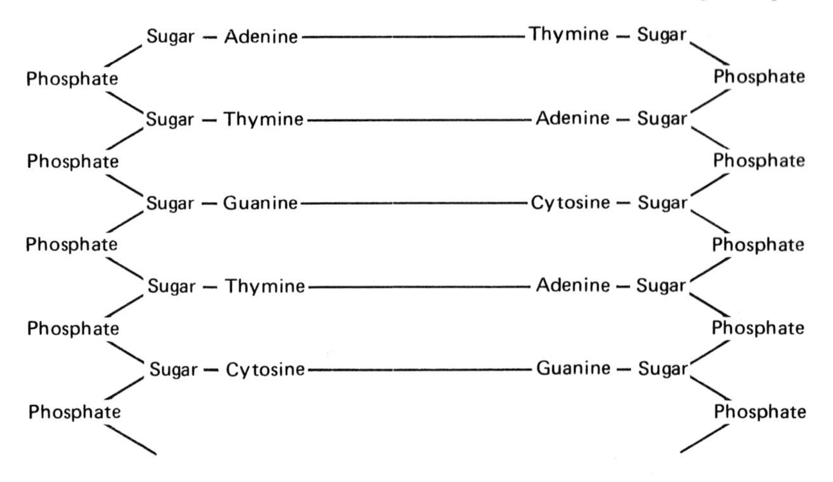

Figure 6.37. Schematic representation of the complementary nature of the two chains in the DNA molecule.

an exact copy of the DNA molecule in the original mother cell. The beauty, and in some respects the simplicity, of the Watson-Crick structure is that each of the polynucleotide chains contains full information to specify the base sequences of the other exactly. Consequently, if the two chains should separate, each can form a template for the synthesis of a new complementary chain. The replication process is illustrated schematically in Figure 6.38. The double-stranded parent molecule separates into two single chains. Each of these chains serves as a template for the synthesis of a complementary chain. Thus, two double-stranded daughter DNA molecules are formed that are indentical in all respects to the parent. This process can of course be repeated for successive generations.

With replication the genetic message has been transmitted from one generation to the next in the four-letter GACT language. Although many of the details of the mechanism of strand separation and of template polymerization still remain to be elucidated, the salient features of replication have been established experimentally. Ingenious isotopic labeling experiments have enabled the progress of the individual chains to be followed through successive generations. The replication process has indeed been found to follow the pathway that has been outlined above. An enzyme that catalyzes the polymerization of DNA, from a pool of the four monomer units, has been isolated by Kornberg. However, for the polymerization to proceed a primer DNA must be present. This DNA presumably acts as a template for the polymerization, since a complementary base composition to the primer is found in the new polymer formed.

In a functioning cell the replicating DNA molecules are organized in a very definite way. They are incorporated into the macrostructure of the individual chromosomes in this form. An electron micrograph of a human chromosome is shown in Figure 6.39. Here the internal arrangement of the DNA molecules with one another is apparent. This collection of DNA molecules is thus highly organized into a superstructural form at levels well above that of the Watson-Crick structure which are

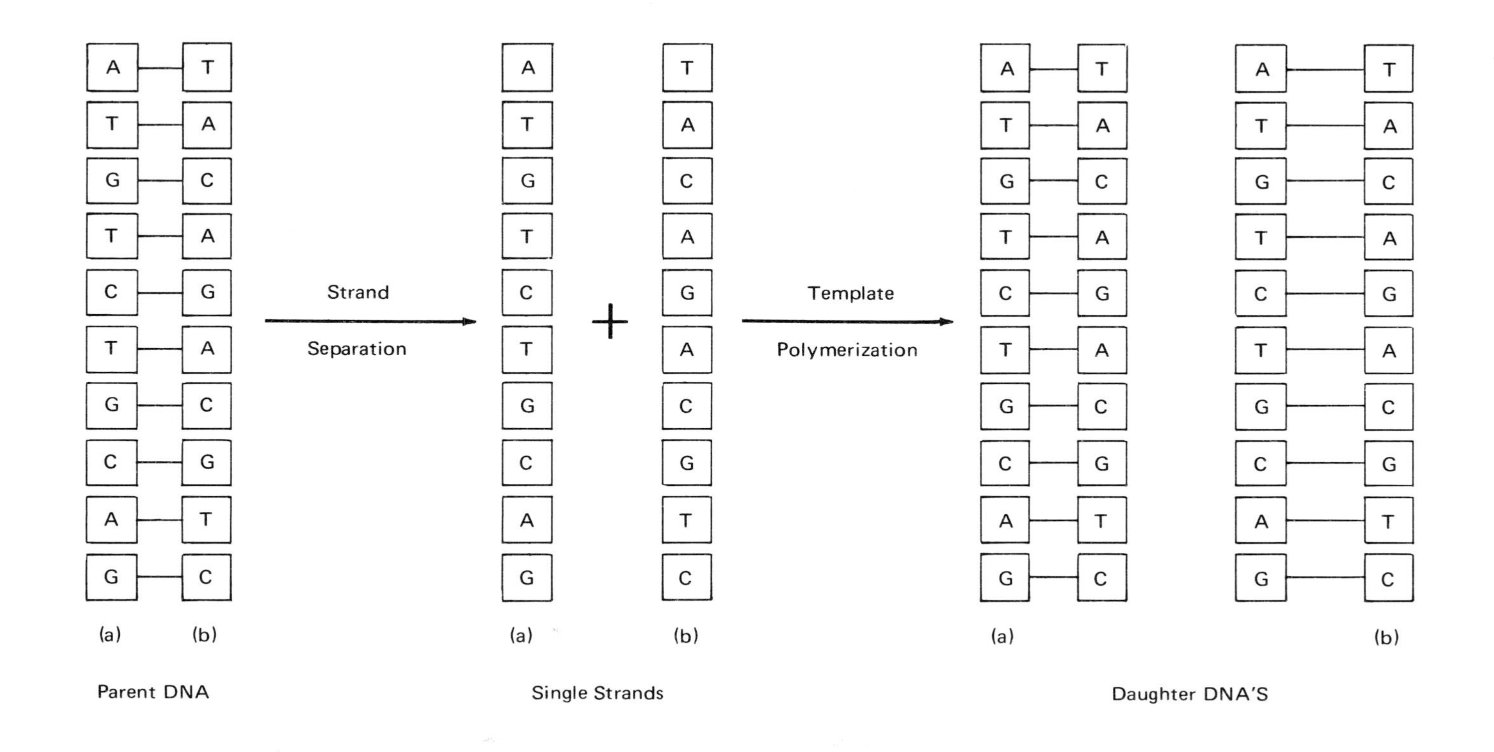

Figure 6.38. Schematic representation of replication process.

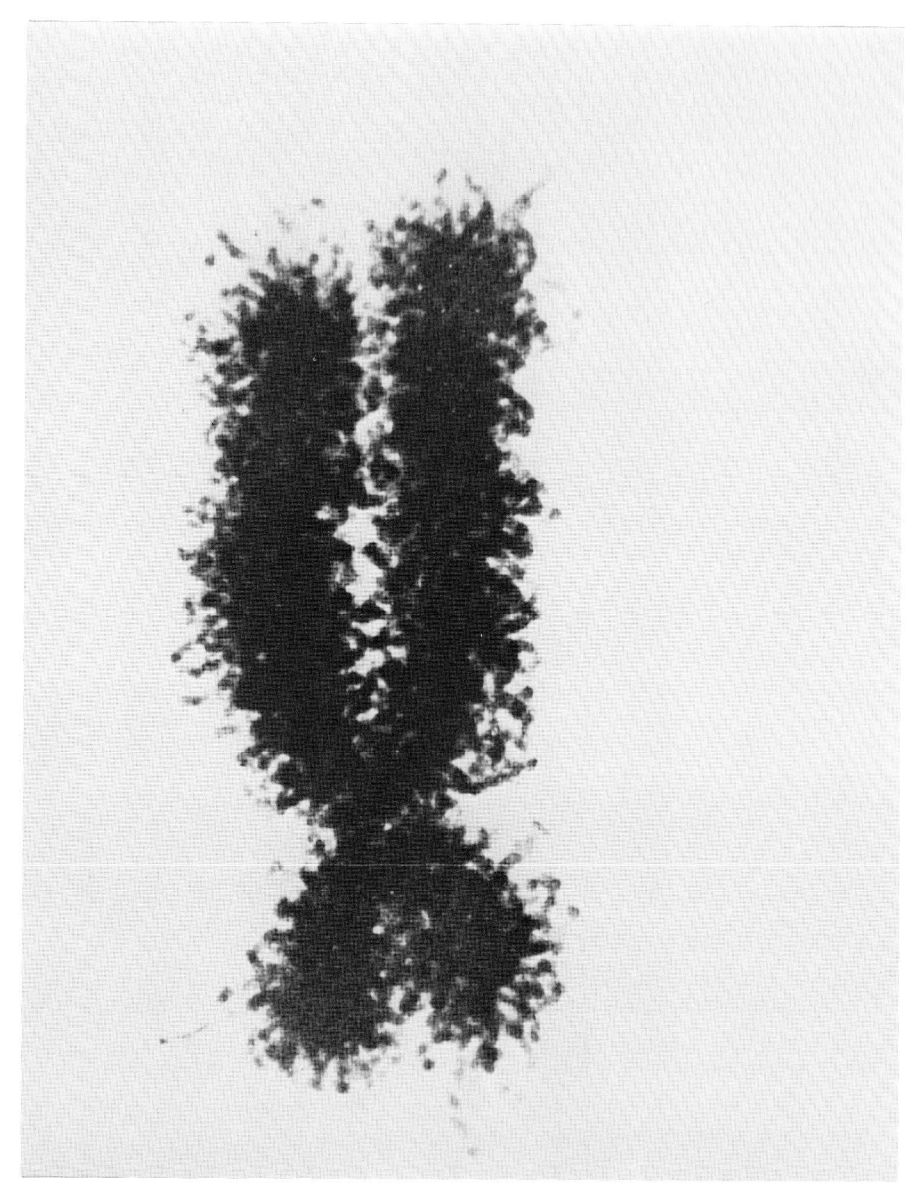

Figure 6.39. A human chromosome magnified 50000 times. From Novikoff and Holtzmann, *Cells and Organelles* 2nd Ed., Holt, Rinehart and Winston. (1976). Reproduced with permission.

maintained locally. The coiling up of the individual molecules in the prescribed manner is aided and abetted by a set of basic proteins called histones, which serve as templates for the nucleic acids.

We have seen how the structure of DNA enables it to perform one of its major functions, that of replication. For an organism to exist it must also have its complement of the many thousands of necessary protein molecules. Although these proteins are all different, they are all synthesized according to instructions contained in the DNA code. Put another way, the DNA molecule contains the information for assembly, in the proper order, of all the amino acids required to make one protein and has this information for all the proteins necessary to sustain living organisms. It has been estimated that the information stored in DNA for a human, if written out in the English language, would require the equivalent of several sets of a 24-volume encyclopedia. Although the information for protein synthesis is contained within the DNA molecule, the extraction of this information and the actual protein synthesis is carried out in the cell through the intervention of the ribonucleic acids, the RNA's.

Ribonucleic Acids–RNA

Before examining the role of RNA in protein synthesis we should recall how it differs chemically from DNA. The sugars in the chain-backbone are different (see Figures 6.29 and 6.32), and in RNA the pyrimidine base thymine (T) is replaced by uracil (U). Three very different kinds of RNA molecules have been identified as being involved in protein synthesis. Each of these is itself synthesized under the control of DNA. These RNAs are called messenger RNA, transfer of t-RNA and ribosomal RNA.

Messenger RNA accounts for about 5 to 10 percent of the total RNA in a cell. It is a short-lived molecule whose function is to receive the code from the DNA of the genes and to carry it to the ribosomes that are in the cell cytoplasm. The ribosomes are the site of protein assembly, or the factory. Thus, messenger RNA is both a transcriber and transmitter of the information for protein synthesis. Transfer RNA is a relatively low molecular weight molecule, composed of from 70 to 80 repeating units. Its function is to attach itself to a particular free amino acid, transport the amino acid to the synthesis site, and to make sure that it is inserted in the proper place in the growing polypeptide chain. Ribosomal RNA, whose molecular weight is on the order of a million, together with certain proteins, form the ribosomes. The ribosomes are small structures scattered throughout the cell cytoplasm. The message and the monomers are brought together here. RNA molecules do not possess the complementary base-pairing characteristics of DNA. They exist as single stranded molecules and except for t-RNA do not exhibit any simple well-defined ordered structure. Yet, despite these apparent shortcomings, these molecules perform their assigned tasks admirably well.

Messenger RNA performs its task because it possesses a sequence of bases that are complementary to a portion of one of the DNA strands. This transcription process, from DNA to RNA, involves the polymerization of the RNA monomers in the correct order utilizing a DNA template. Within the cell nucleus, a portion of the DNA molecule unwinds and serves as a template for the polymerization of messenger RNA. The polymerization requires the active help of a specific enzyme. For example, a DNA base sequence of A–T–C–G–T will be transcribed into U–A–G–C–A in a messenger RNA. The sequence of bases in DNA determines the sequence of bases in RNA. In the transcribed molecule the bases control the order in which the amino acids are incorporated into a growing polypeptide chain. This is the genetic code in action. The question then arises as to what specifically is the message.

The operation of the genetic code must obviously involve a relationship between the four bases on the nucleic acids and the 20 possible amino acids used in making a protein. Since there are only four different bases there cannot be a one to one correspondence between a base and an amino acid. Similarly, since there are at most 16 different nucleotide pairs in sequence, a direct correspondence between pairs of nucleotides and all amino acids must also be ruled out. The coding for a single amino acid residue must, therefore, contain at least a sequence of three nucleotides. It turns out that the code is a triplet or three-letter word such as AUC. Since there are 64 combinations of triplets (4^3), there must be a redundancy, or degeneracy as it is called, in the coding. Before examining the specifics of the genetic code we summarize in Figure 6.40 the role of DNA and messenger RNA in the storage and transcription process. The necessary information is stored in the two stranded base-paired DNA molecule in the sequence of bases. The message is transscribed onto messenger RNA using the first strand of the DNA. Each successive grouping of three bases on the messenger RNA, called a codon, codes for the particular amino acid indicated. In this example, we have actually shown the last 10 amino acid residues of ribonuclease (see Figure 6.12). A sequence of repeating units in the DNA, i.e., a portion of the complete molecule, that codes for a particular protein is said to direct its synthesis and is called a gene.

In a set of ingenious experiments from the laboratories of Nirenberg, Ochoa, and Khorana, which will have far-reaching implications for mankind, the direct correspondence between the base triplets and the amino acid residues has been established. The results of these experiments are summarized in Table 6.3. All 64 possible triplet combinations are accounted for. Some of the amino acids are coded for by as many as six different triplet combinations. On the other hand, methionine and tryptophan are each coded by single combinations AUG and UGG respectively. Also included in the table are the code words for the termination of the growing chain, i.e., the instruction to stop the polymerization when the protein assembly has been completed. We note that there is no code word for two of the amino acid residues found in proteins–hydroxyproline and cystine. These residues must obviously be synthesized from their close relatives, proline and cysteine, after the polypeptide chain is formed. Thus, the covalent disulfide cross-links typical of many proteins, and involving residues well removed from one another along the chain, are formed after the protein is put together.

Before discussing protein synthesis further, we shall examine the consequences of mistakes in the primary code of the DNA or in its transcription. An error in the

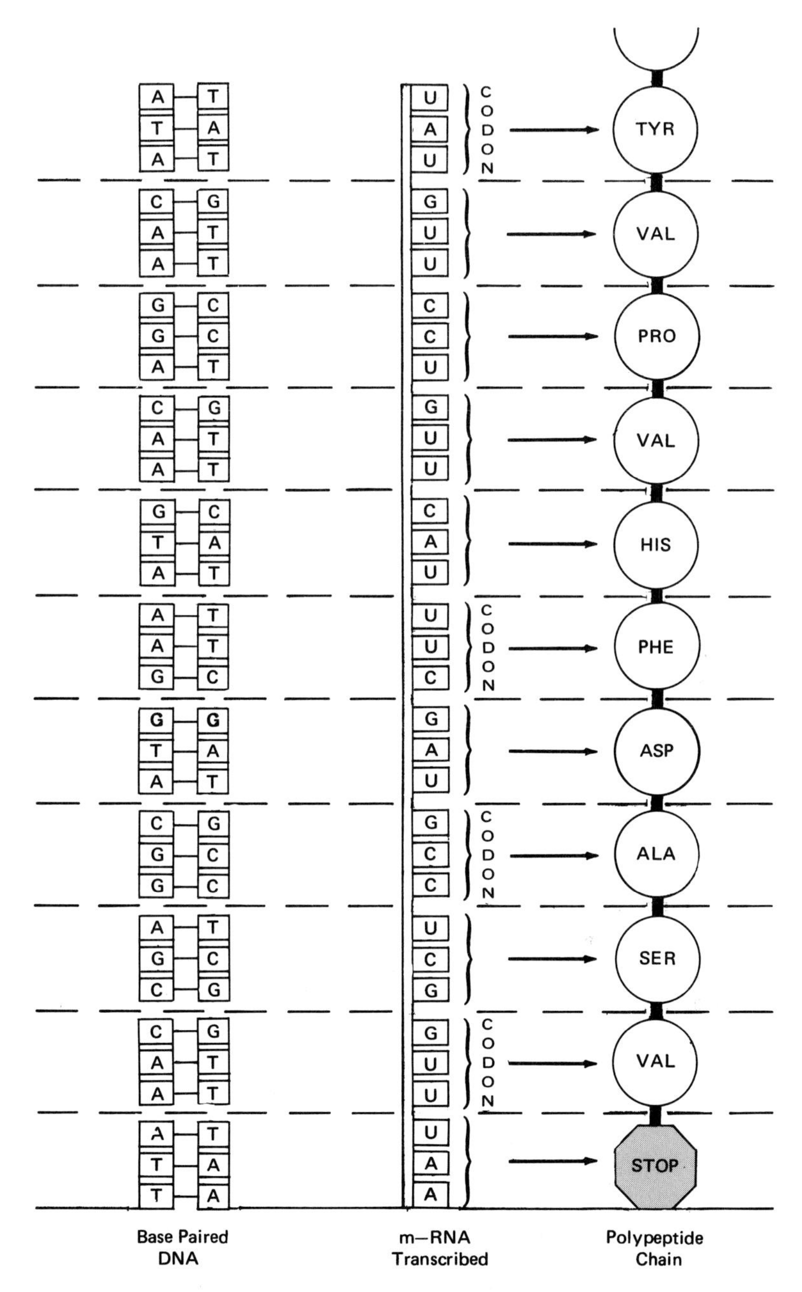

Figure 6.40. Schematic diagram of storage and transcription role of the nucleic acids in directing syntheses of the last ten residues of ribonuclease.

Table 6.3. Genetic Code: Relation Between Base Triplets and Amino Acids

Amino Acid	Triplet or Codon Assignment					
Alanine	GCC	GCU	GCA	GCG		
Arginine	CGC	CGU	CGA	CGG	AGA	AGG
Asparagine	AAU	AAC				
Aspartic Acid	GAU	GAC				
Cysteine	UGU	UGC				
Glutamic Acid	GAA	GAG				
Glutamine	CAA	CAG				
Glycine	GGU	GGC	GGA	GGG		
Histidine	CAU	CAC				
Isoleucine	AUU	AUC	AUA			
Leucine	UUA	UUG	CUU	CUC	CUA	CUG
Lysine	AAA	AAG				
Methionine	AUG					
Phenylalanine	UUU	UUC				
Proline	CCC	CCU	CCA	CCG		
Serine	UCU	UCC	UCA	UCG	AGU	AGC
Threonine	ACU	ACC	ACA	ACG		
Tryptophane	UGG					
Tryosine	UAU	UAC				
Valine	GUU	GUC	GUA	GUG		
Terminate (Stop)	UAA	UAG	UGA			

position of one of the bases, i.e., if it were incorrect or misplaced, results in a different triplet, so that a different amino acid residue in a particular position in the protein results. These changes are usually brought about accidentally in nature by the action of high energy radiation or by chemical agents. Because of these errors, changes or mutations are found in the proteins of the progeny. In some cases these mutations have only a minor effect, but in others they can be disastrous. As an example we have already noted that in sickle cell anemia the glutamic acid residue in the sixth position in the β-chain of hemoglobin is replaced by valine. As Table 6.3 shows, the codons for glutamic acid are GAA and GAG. If the middle base is replaced by U, the resulting triplets GUA or GUG code for valine. We see then that this apparent superficially small change is the genetic basis for the particular molecular disease. Its cure must involve correcting the DNA message. Sometimes the message is so garbled in a particular gene that an important enzyme cannot be synthesized. This can cause a major disruption of metabolic processes and very often leads to early death or mental retardation among infants.

Returning to the problem of protein synthesis, we have already pointed out that the monomers are brought to the polymerization site by t-RNA. The transfer RNA molecule must not only select the proper amino acid from the monomeric pool but must also be able to read the code on the messenger RNA so that the amino acid is placed in the proper position in the growing chain. All the t-RNA molecules contain about 70 to 80 repeating units. In addition to the usual RNA bases, other are found in small but significant proportions. There is at least one particular t-RNA for each amino acid; different amino acids do not become attached to the same t-RNA molecule. In some cases, however, each of several different t-RNA molecules can select and attach itself to the same amino acid. Because of the highly specific function of t-RNA we would anticipate that the molecule would have a very particular structure designed to carry out its responsibilities.

Several of the t-RNAs have now been purified and the detailed sequence of their bases has been determined. The three-dimensional structure of one of the t-RNAs, that for phenylalanine, has now also been established in detail. Hence, we can now deduce the broad outline of how these molecules function by correlating their chemical and spatial structures with the specific acts that they perform. All transfer RNAs have a common sequence of terminal bases, namely C–C–A. The sugar (ribose) unit associated with the terminal adenine base has both of its hydroxyl (OH) groups free. As is shown in Figure 6.41, the amino acid is attached to one of the OH groups. The specificity of amino acid selection involves the intervention of an enzyme particular to the amino acid. The molecule must also have a sequence of three bases that are not hydrogen-bonded to any other and that are complementary

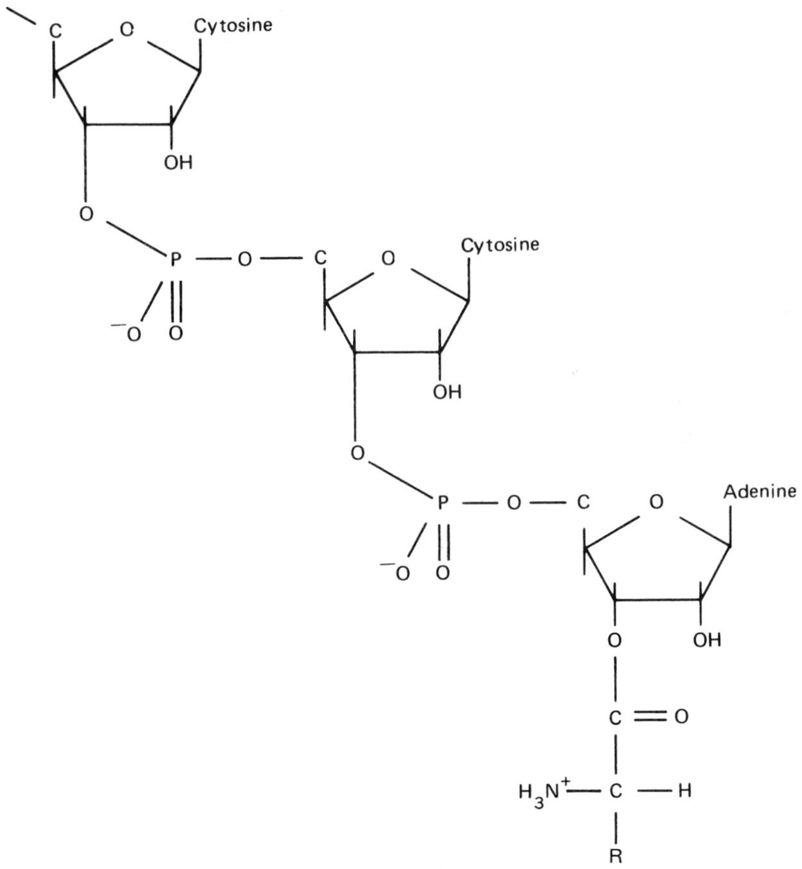

Figure 6.41. Terminal bases of t-RNA.

to the condon of the messenger RNA. This structural feature will allow the amino acid to be inserted in the proper place. For example, if methionine is to be transferred, then the codon on the messenger RNA calling for this amino acid is AUG. Therefore, the complementary sequence, or anticodon, UAC capable of hydrogen bonding with AUG must be present in the t-RNA. In this way the t-RNA will find its proper match along the messenger RNA. The minimal structural requirements of this t-RNA are represented in the highly schematic cloverleaf form in Figure 6.42.

Based on an analysis of the data available for the phenylalanine t-RNA, the actual three-dimensional structure turns out to be in the shape of an extended L, as is represented in Figure 6.43. One arm of this L-shaped molecule contains the terminal of acceptor end. The other arm is formed by the anti-codon sequence or loop. The specificity of the base sequences along this nucleic acid molecule determines its three-dimensional structure. This structure is uniquely designed to carry out the molecules assigned function.

In the process of protein synthesis, when the t-RNA has acquired its amino acid, if diffuses to the ribosomes. Here, after appropriate interaction with the messenger, RNA synthesis takes place. The ribosomes are roughly spherical particles about 200 Å in diameter. They are composed of both RNA and protein and have a molecular weight of about 2.7×10^6. Ribosomal RNA does not contain genetic information, and the details of its role in protein synthesis are still obscure. The ribosome is composed of two distinct subunits one being about twice the size of the other, as it shown in Figure 6.44. The protein synthesis does not take place in homogeneous

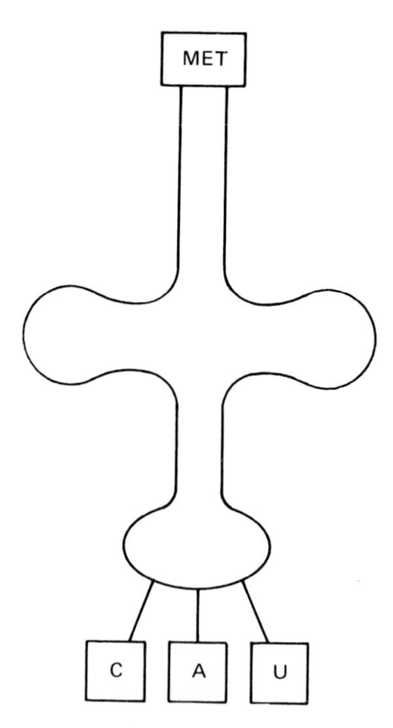

Figure 6.42. Schematic structure of methionine t-RNA.

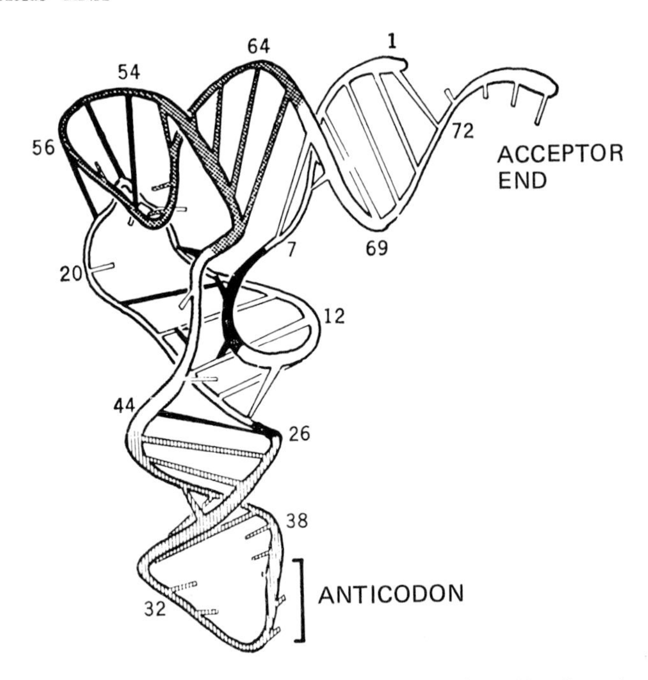

Figure 6.43. Schematic three-dimensional structure of t-RNA for phenylalanine. Numbers represent the nucleotide residues. Adapted from S. H. Kim *et al* Science *185* 435 (1974). Reproduced with permission. Copyright (1974) by the American Association for the Advancement of Science.

solution but is carried out on the surface of the ribosomes. The chief function of the ribosome is to orient the t-RNA and its associated amino acid with the messenger RNA so that the code can be read accurately. Ribosomes must thus contain specific surfaces that bind t-RNA, messenger RNA, and the growing polypeptide chain in proper orientation.

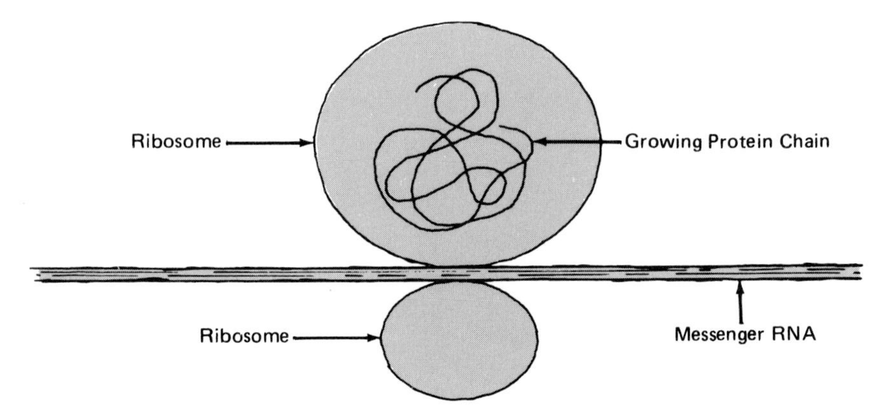

Figure 6.44. Schematic representation of ribosome interacting with messenger RNA.

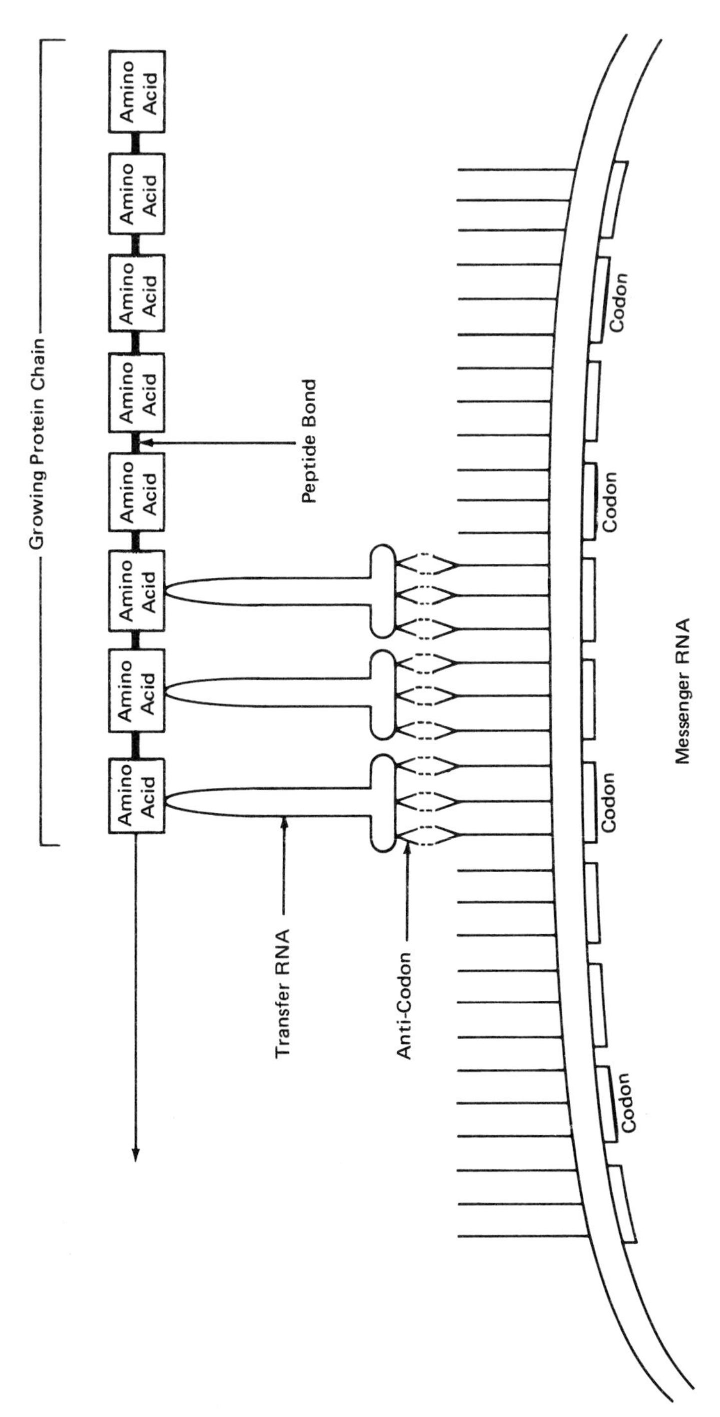

Figure 6.45. Schematic representation of protein synthesis.

A schematic representation of the polymerization is given in Figure 6.45. The t-RNA and the growing polymer chain appear to be attached to the larger of the ribosomal subunits; whereas messenger RNA is bound to the smaller one. The growing chain is retained by the ribosome until synthesis is complete. There is a great deal still to be understood about details of protein synthesis and the specific role played by the ribosomes. It is known, however, that the polypeptide chain is initiated and grows from the N-terminal amino acid end in a stepwise manner. As each repeating unit is added, the ribosomes move in a fixed direction relative to the messenger RNA. This motion allows the next codon to interact with a t-RNA. The amino acid at the other end of the t-RNA is thus in position to be properly attached to the peptide chain. After the "stop" codon is reached, the chain is released by the ribosome. The sequence of amino acid residues determines the proper three-dimensional structure of the protein. The time required for the synthesis of a polypeptide chain containing 300 to 400 amino acid residues is about 10 to 20 seconds.

In a major scientific and technological development, biochemists and molecular biologists have developed techniques that enable a particular gene, from the multitudinous number available, to be extracted or excised from the DNA of a human or animal cell and to be inserted into the DNA of a bacterium. The skillfull use of several very particular sets of enzymes are involved in this process. The new molecule that is thus formed is called a recombinant DNA. As the bacteria multiply they will now make millions of copies, not only of their own genes but also of the foreign

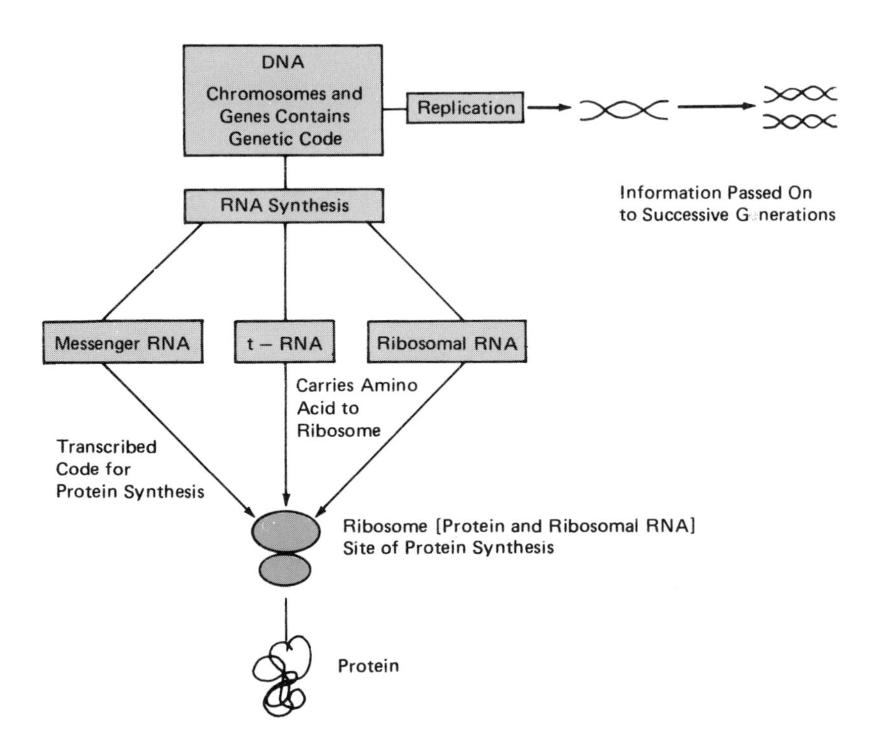

Figure 6.46. Summary diagram of role of nucleic acids in biological function.

gene that has been inserted into its DNA. Thus, a molecular cloning process has been developed that provides a way for the propagation of genetically similar organisms. These all contain aritificial, but identical, DNA molecules. It is now possible, therefore, to transfer genetic information from one kind of organism to a completely unrelated one, which can then replicate itself.

As we can recall, the instructions for the natural synthesis of a protein is encoded in the DNA sequence of a corresponding gene. Now as the bacteria multiply they not only make copies of their own genes but also of the foreign one that has been inserted within it. Concomitantly, and most importantly, the bacteria will also synthesize the protein specified by the foreign gene. Thus, a very efficient factory has been constructed for the production in relatively large quantities of a particular protein. In natural synthesis, and in the few rare laboratory procedures available, only an extremely small quantity of protein can be prepared. Examples of proteins and peptides already synthesized by the recombinant technique include insulin, human growth hormone, and interferon, which is an anti-viral and perhaps anti-cancer agent and a vaccine against foot and mouth disease that is common among livestock. This new technology will clearly have far-reaching consequences, the potential of which is just beginning to be understood. Besides producing theraupeutic agents, enzymes will be synthesized that will be used to perform specific chemical reactions and produce compounds such as ethanol and methane. Application to agriculture and food processing are easily envisaged. Entirely new industries are going to be developed based on this new technology.

Before concluding we summarize, by means of Figure 6.46, the diverse roles of the nucleic acids in biological function.

EPILOGUE: MACROMOLECULES AND MAN

We have seen from our discussion of macromolecular structure and properties the features that are common to these substances as well as the diversity of functions they perform and the many uses to which they are put. Their essential nature to life as we know it today, both as a biological necessity as well as to our comfort and well being, is readily apparent. With the fundamental understanding that has evolved regarding this class of molecules we can anticipate, with great confidence, the development of whole new classes of polymeric substances as well as the laboratory synthesis of many biopolymers that are crucial to life processes. The history of science has taught us to expect that much more will yet be coming. The present, as well as the anticipated advances in polymer science, will have a profound impact on our society as it influences our standard of living, our ecology and environment and through molecular biology and molecular biophysics, our control of biological processes. This ever-increasing scientific bounty will bring with it many problems and responsibilities. The solution of these problems and the proper and conscientious assumption of these responsibilities rests with all of society. However, intelligent solutions cannot be devised without a real scientific understanding of the molecular nature of the systems with which we are concerned. If we have accomplished this latter task in this book, then it is appropriate to outline and consider briefly the major problems that are already with us or that can be easily anticipated in the very near future.

Synthetic materials, particularly plastics, are often thought of as being cheap, inferior products of our highly technological Western society. It is argued that we can easily do without such materials. In an absolute sense this may be true for a limited number of products. In a relative sense, however, this concept is clearly not warranted. The properties of most synthetic materials equal or surpass those provided by nature. If we were to depend solely on natural sources for these materials they would soon become depleted because of the demand, even if the earth's population remained constant. This depletion would have devastating effects on our environment since our forests would be stripped and our land eroded. Those of us who live in affluent societies where we take our shelter, clothing, and transportation for granted could undoubtedly very easily dispense with a great deal of our new materials and luxury. This is not so, however, for those who live in poorer societies. More than half of the earth's population is made up of the inhabitants of the underdeveloped countries of the world. For those people, what we consider nonessentials can easily serve as a base for a substantial improvement in the quality of their life. Here the use of substitute synthetic materials offers by far the greatest opportunity to enhance their standard of living and bring it nearer to a level that we have come to expect and enjoy. By taking advantage of the technological advances that have been made in polymer science and of the large quantities of polymeric materials that can be produced very economically, they can be provided with many of the comforts of life.

For example, when one starts virtually from scratch without concern for traditional methods of production and vested interests, then housing and clothing and tires for transportation can be produced economically and can be widely distributed. Poly-*cis*-isoprene, natural rubber, can now be manufactured in chemical plants. The plantation-based production and economy can be eliminated and the commercial products derived from this polymer could be made more widely available. Synthetic fiber plants, strategically situated around the world, could supply clothing for all. Major advances can also be made in housing. The use of polymer-fiber glass reinforced materials or polyurethane foam for the housing shells makes possible sturdy but inexpensive and rapid construction. This concept in housing is not a theoretical dream but is actually being realized in a very modest way today. This effort only has to be expanded to bring decent shelter to everyone. Furnishings made from the ABS type copolymers can make living more pleasant and comfortable. The use of polymeric substances as semiconductors and their continued and more expanded use in micro-electronic devices can be anticipated. In the area of health care and needs, we find the use of synthetic polymers becoming increasingly important as major components of artificial organs, as for example the heart. Polymers are also finding extensive use in a wide variety of prosthetic devices. A whole new field of biomaterials has developed into maturity. The ability to deliver drugs to specific organs and tissues and the development of slow-release systems are further examples of the application of polymer science to human needs. The same basic science and technology, that we have been discussing in detail, is applicable to all these cases. These few examples indicate the comforts and health care that can be provided that would otherwise be virtually unattainable. Polymer technology should thus be directed to the benefits of humanity so that its great promise will be fulfilled. It is our responsibility as members of society to make sure that such efforts are vigorously pursued.

There are, then, very compelling reasons for the production of synthetic materials and the continued research and quest for new and better products. If we accept this principle, then we must also realize that it presents a very serious problem and challenge. The problem in its simplest form is that of waste disposal. Although many naturally occurring polymers such as cellulose and natural rubber will degrade with time, i.e., are biodegradable, synthetic materials in the form of fibers, containers, and packaging material must be disposed of. We certainly do not want a major portion of the earth surface to become covered with polymeric debris. Unlimited production must eventually lead to this situation, since polyolefins such as polyethylene and polypropylene, and vinyl polymers such as polystyrene and polyvinyl chloride, are not biodegradable. They cannot be recycled as paper, glass, and certain metal containers can, nor does the development of recycling processes appear technologically feasible in the foreseeable future. Is the solution to this vexing problem the limitation of production either indirectly by taxation, as has already been attempted in some cities, or directly by legal restrictions? If so, this obvious and very simple approach would mean denying the benefits of the technological advances in substitute and very utilitarian material to those segments of our civilization whose standard of living is low and can be easily raised. This dilemma can be resolved if we take advantage of the fact that most of the polymers will burn. What is required to solve the problem, then, is the development of incinerators that are self-contained and thus will not cause pollution of the atmosphere. Although such incinerators have not been developed they are certainly technically feasible and represent a way to protect our environment while still making available the benefits of new materials to the people who need them the most. Much effort should thus be given to the development of such incinerators. Serious problems do not usually have simple solutions. We must be concerned with the impact on all of civilization rather than the small segment of society in which we find ourselves when we decide on a course of action.

Another set of very serious responsibilities faces us in the light of the revolution in molecular biology that has occurred in our lifetime. Many far-reaching implications are inherent in the discoveries that have already been made. For the first time in his history man understands his chemical origin and function. He can now investigate and understand life at its most fundamental molecular level and has the capability, in principle, to control the design and fate of future generations. DNA has already been synthesized in the laboratory, as have certain genes and a completely functioning enzyme. The development of the recombinant DNA technology gives virtually unlimited promise for the production of drugs, in application to agriculture and indsutry, and most important of all for the further and better understanding of biological processes and molecular biology. In the not too distant future, man will have the capability to create man.

The promises to mankind that can be derived from this knowledge are very great. Cure and control of genetic diseases and detection and rectification of genetic errors so that our progeny will not be abnormal or deformed, all lie within our grasp. The molecular structure of viruses and their control and relation to cancer and other diseases will be unraveled before long. These benefits, coming as they will from our understanding of the structure and function of biological macromolecules, justifies the effort, the time, and frustration of the basic scientific research that underlies all these efforts.

Accompanying all the good that we can foresee there are also many danger signals. It is no longer in the realm of science fiction to anticipate the creation of life in a test tube and the synthesis of living organisms. Is this to go on uncontrolled, left to individuals' desires and passions? Major experimental efforts designed to understand brain function, memory, and the learning process are now under way in many laboratories. Proteins and nucleic acids are involved in these functions, and we can expect a fundamental understanding of these processes to emerge shortly. Clearly these are no longer scientific or technological problems but problems that must be the deep concern of society as a whole. How we address ourselves to these matters and control our actions may very well decide whether man will survive or become another extinct species in evolutionary history. It is not too early for us to become deeply involved and concerned.

The acceptance of the macromolecular hypothesis and the consequences of scientific study have given us materials unthought of 40 years ago, as well as a detailed understanding of the fundamentals of life processes. We stand on the threshold of a new era in our civilization. Whether the Age of Macromolecules will be man's golden age or a disastrous period clearly rests with man himself.

SUGGESTED BIBLIOGRAPHY FOR ADVANCED READING

For a more detailed discussion of polymer properties, at about the same level as this monograph, consult

R. B. Seymour, *Introduction to Polymer Chemistry,* McGraw-Hill Book Company, New York.

Some suggested intermediate level texts are

R. Allcock and F. W. Lampe, *Contemporary Polymer Chemistry,* Prentice-Hall, Inc. Englewood, New Jersey.

F. A. Bovey and F. H. Winslow (eds.), *Macromolecules: An Introduction to Polymer Science,* Academic Press, New York.

The classic treatise in the field, which is a superb exposition, is

P. J. Flory, *Principles of Polymer Chemistry,* Cornell University Press, Ithaca, N.Y.

This book is very strongly recommended for those who wish to pursue the study of polymers in more detail. A knowledge of undergraduate physical chemistry is, however, necessary for a full understanding.

There are several monographs on specialized subjects. They can be read with the background that has been developed here. These are

L. R. G. Treloar, *Physics of Rubber Elasticity,* Oxford University Press, Oxford.

L. Mandelkern, *Crystallization of Polymers,* McGraw-Hill Book Company, New York.

J. D. Ferry, *Viscoelastic Properties of Polymers,* John Wiley and Son, New York.

R. W. Lenz, *Organic Chemistry of Synthetic High Polymers,* Interscience Publishers, New York.

In the area of biopolymer structure and the properties and function of proteins and nucleic acids, further suggested reading can be found among the following

G. E. Schulz and R. H. Shrimer, *Principles of Protein Structure,* Springer-Verlag, New York.

R. G. Dickerson and I. Geis, *The Structure and Action of Proteins,* Harper and Row, New York.

J. D. Watson, *Molecular Biology of the Gene,* Third Edition, W. A. Benjamin, Inc., Menlo Park, Ca.

E. J. DuPraw, *Cell and Molecular Biology,* Academic Press, New York.

There are also films available. These are:

Physical Chemistry of Polymers, prepared by Bell Telephone Laboratory. Available from Sterling Movies Inc. 43 West 61st Street, New York, New York 10023.

Chemistry of the Cell. I: The Structure of Proteins and Nucleic Acids II. Function of DNA and RNA in Protein Synthesis. Available from the McGraw-Hill Book Company. New York.

INDEX

ABS, 17, 18, 154
Actin, 120
Acrylic fiber, 18
Adenine, 132, 134, 147
Adenosine triphosphate (ATP), 98
Albumin, 122
  egg, 112
Allcock, R., 156
Allen, R.C., 82
Amide bond, 104
Amide group, 105
Amino acids, 100, 101, 102, 103, 109, 111, 112, 113, 114, 116, 117, 122, 123, 127, 143, 146
Amylopectin, 13, 17
Amylose, 12, 13
Anemia
  ferrihemoglobinemia, 129
  hemolytic, 129
  sickle cell, 129, 146
  thalassemia, 129
Anfinsen, C.B., 107, 137
Atactic, 50, 51
Anti-codon, 148, 149, 150
Aramid, 11, 15, 88, 92, 93, 94
Arteries, 111
Artificial organs, 154
Asymmetric, 43, 53, 89, 91

Baekeland, L., 24
Bakelite, 24
Biomaterials, 154
Bone, 117
Bovey, F.A., 156
Branches
  long-chain, 4
  short-chain, 5
Branching, 4, 24, 85, 87
Brown, C.J., 44, 71
Bunn, C.W., 44, 45, 48, 69, 70, 71
Butadiene-styrene copolymers, 17
n-Butane, 14, 31, 32, 33
Butyl rubber, 10, 65, 88

Carbohydrates, 1
Cartilage, 117
Cellophane, 12
Celluloid, 12, 24
Cellulose, 12, 13, 15, 16, 17, 21, 89, 93
  derivatives, 86
Cellulose acetate, 12, 89
Cellulose nitrate, 12, 24
Cellulose triacetate, 13, 88
Cellulose trinitrate, 13, 86, 88
Chain dimensions
  average, 36, 37, 38, 39, 40
Chain mobility, 66
Chargaff, E., 134
Chiang, R., 84
Chick, F.H.C., 134, 135, 140
Chromosomes, 130, 134, 140, 142
Claws, 111
Cloning, 152
Coatings, 1, 25
Codon, 145, 148, 150, 151
Collagen, 26, 73, 74, 94, 96, 112, 113, 118, 119, 120
Compressibility, 63
Conformation
  definition, 28
  ethane, 29
  eclipsed, 29, 30, 31
  gauche, 31, 32
  general, 34, 53, 68, 70
  helical, 47, 49, 50, 52, 78
  ordered, 110, 118
  planar zig-zag, 42, 48, 50, 68, 78
  staggered, 29, 30, 31
  statistical, 34, 35, 40, 57, 62, 94, 122
  trans, 31, 32
Connective tissue, 111, 114
Contractility, 94, 95, 97, 99, 111
Copolymers, 4, 7, 17, 19
  alternating 7, 8, 17
  block, 7, 17, 85, 86, 116
  graft, 17
  ordered, 7
  random, 7, 17, 19, 85, 86

Copolymers (*cont.*)
stereo-irregular, 50
Corey, R.B., 105, 109, 114
Cornea, 117
Corradini, P., 108
Cotton, 1, 12, 91, 93
Cross-linked, 5, 7, 24, 26, 59, 95, 114, 115
Crystalline state
polymers, 50, 53, 68, 72, 79, 89
monomers, 72
Crystallite, 72, 73, 75, 76, 78, 79, 85, 93
Crystallization process, 70, 72
Cysteine, 144
Cystine, 101, 109, 114, 144
Cytosine, 132, 134, 147

Dacron, 11, 88, 89, 91, 92
Danusso, F., 52
Daubeny, R.F., 44, 71
Debye, P.W., 28
Delrin, 10, 88
Density, 80, 81, 83, 94
Deoxy-D-ribose, 131, 132
Diamond, 93
Dickerson, R.E., 124, 126, 130, 156
DNA (see nucleic acids)
Diseases, 129
foot and mouth, 152
Drawing, 73, 90
D-Ribose, 131, 132
DuPraw, E.J., 136, 138, 156
Dynel, 18

Ecology, 153
Elastic properties
deformation, 54, 59
modulus, 61, 64, 81, 83, 91, 93
rubber, 54, 57, 62
Elastomer, 17, 18, 19, 54, 68, 86
Elvax, 18
Electron microscopy, 75, 76, 77, 78, 79, 80, 81, 114, 115, 117, 118, 120, 121, 135, 138, 139, 140
Encapsulating, 19, 25
End-to-end distance, 34, 36, 39, 58, 59
Engineering plastics, 14, 63, 66, 87
Entropy, 56, 57
Enzymes, 16, 111, 123, 124, 127, 140, 147, 152
Epidermin, 114
Epoxies, 25, 26
Ethane, 13, 29, 30, 32
Ethylene, 21, 31
Ethylene-propylene copolymer, 17

Ferry, J.D., 156
Fibers, 17, 43, 68, 69, 86, 88, 89, 90, 92, 94, 95, 116, 154
Fibrin, 114
Fibrinogen, 122
Fibroin, 112
Filshie, B.K., 115
Fischer, E.W., 80, 81
Fleming, A. 123
Flory, P.J., 23, 28, 33, 37, 42, 60, 84, 95, 96, 97, 156
Free radical, 22, 26
Functionality of reaction
bifunctional, 20, 24, 104, 134
general, 19
monofunctional, 20
trifunctional, 24

Gamma rays, 26
Geis, I., 124, 126, 130, 156
Gelatin, 117
Gene, 130, 144, 151, 152
Genetics
code, 134, 143, 144, 146, 151
disease, 146, 155
information, 2, 27, 130, 131, 134, 143, 148, 152
Geometric isomers, 68, 85
Glasses, 93
Glass formation, 61, 65
Glass temperature, 62, 65, 66, 67, 69
Globulin, 122
Glucose, 16, 17
Goodyear, C., 26, 59
Gough, J., 54, 55
Graft polymers, 8
Grobstein, C., 139
Growth hormone, 152
Guanine, 132, 134
Gutta percha, 10, 13, 15, 88

Hair, 26, 111, 114, 115
Harwick, E.R., 122
Heat of fusion, 80, 81, 83

α-Helix, 42, 105, 106, 107, 108, 109, 114, 123, 128
Heme group, 127, 128
Hemoglobin, 2, 111, 122, 127, 128, 129, 130
Hexane, 14
Hide, 114, 117
Holmes, D.R., 45
Holtzmann, J., 142
Holum, J.R., 119
Homopolymers, 4, 8, 13, 15
Hooke's law, 58, 61, 88
Howells, E.R., 48
Hydrogen bonding, 44, 69, 106, 117, 135, 147
Hydrophobic, 101, 122, 123
Hydrophyllic, 101, 122, 123
Hydroxyapatite, 117
Hydroxyproline, 101, 117, 118, 144

Inorganic polymers, 12, 13
Insulin, 112, 152
Interface, 76, 79, 85
Interferon, 152
Internal, 28, 29, 32
Ionizing radiation, 26
Isotactic, 41, 50, 51, 52, 85, 106

Kendrew, J.C., 128
Keratin, 112, 114, 115, 116
Kevlar, 11, 88
Kim, S.H., 149
Khorana, H.G., 144
Knobler, C.M., 122

Lampe, R.W., 156
Leather, 26
Ligaments, 117
Linear homopolymer, 1, 4
Lipoprotein, 122
Liquid-crystal, 43, 71, 73, 89, 91
Lucas, F., 116
Lucite, 9
Lysozyme, 123, 124, 125, 127

Mandelkern, L., 74, 75, 80, 81, 156
Marsh, R.E., 109
Mechanical properties, 62, 81, 88, 89, 91, 92
Mechanochemistry, 94
Melamine, 25
Melting process, 82, 85
Melting temperature, 84, 85, 86, 87, 92, 96
Membranes, 111
Methane, 13
Microelectronic, 154
Mitosis, 99
Mitotic apparatus, 111
Molecular biology, 153
Molecular biophysics, 153
Molecular sizes, 20
Molecular weight, 23
  average, 21, 23
  distribution, 23
Morphology, 73, 75, 78, 79, 85, 91
Motility, 99, 121
Murray, J.M., 121
Muscle, 111, 117
  contraction, 99, 114
  fiber, 98, 120, 121
Myoglobin, 127, 129
Myosin, 120

Nails, 111
Natta, G., 50, 108
Natural rubber, 1, 10, 13, 15, 42, 59, 63, 65, 68, 70, 73, 74, 86, 88, 94, 95, 154
Nematic state, 71
Neoprene, 10, 65
Network
  deformation, 59, 60
  formation, 24
  infinite, 5
  three-dimensional, 5, 6, 24, 59, 126
  structure, 60
Neutrons, 26
Nirenberg, M., 143
Nomex, 11
Novikoff, H., 142
Nucleation, 72
Nucleic acids
  deoxyribonucleic acid (DNA), 131, 133, 138, 139, 140, 143, 158
  general, 1, 2, 13, 27, 42, 100, 130, 131, 151
  messenger RNA, 143, 148, 149, 150, 151
  ordered structure, 135, 136, 140
  ribonucleic acids (RNA), 131, 133, 143

Nucleic acids (*cont.*)
ribosomal RNA, 143, 151
transfer (t) RNA, 143, 146, 147, 148, 149, 151
Nylon 6, 11, 44, 45, 91, 92, 94
Nylon 6-6, 10, 44, 45, 88, 89, 91, 92, 93, 94

Ochoa, S., 144
Orlon, 9, 18
Oth, J.F.M., 95

Pauling, L., 105, 109, 114
Perutz, M.F., 128
α-Phenylene polyamide, 42, 43, 46, 88, 93
Plasticizers, 66
Plexiglas, 9, 62
Polyacetylene, 11
Polyacrylonitrile, 9, 18, 86, 88, 89
Polyadenylic acid, 13
Polyalanine, 13, 105
Poly-N-alkyl isocyanate, 42
Polyamides, 15, 21, 40, 43, 44, 54, 69, 86, 91, 92, 108
Poly α-amino acid, 13, 40, 41
Poly p-benzamide, 11
Poly bisphenol-A-carbonate, 65, 88
Polybutadiene, 10, 65
Polycaprolactam, 11
Polycarbonate, 11, 14, 87
Polychloroprene, 10, 65
Polydichloro phosphazene, 12, 65
Polydimethyl siloxane, 10, 41, 65, 86, 88
Polyester, 11, 15, 40, 43, 44, 54, 69
Polyether, 10, 11
Polyether sulfone, 65
Polyethylene, 2, 4, 5, 9, 11, 14, 17, 19, 26, 28, 29, 34, 36, 40, 42, 44, 68, 69, 72, 73, 74, 75, 77, 79, 80, 81, 83, 85, 86, 87, 88, 92, 95, 97,
branched, 4, 6, 26
Polyethylene adipate, 11
Polyethylene oxide, 10, 41, 82
Polyethylene terephthalate, 11, 44, 65, 69, 71, 86, 88, 89, 91, 92
Polyglycine, 40, 41, 105
Polyhexamethylene adipamide, 10, 41, 65, 86, 88, 89
Polyisobutylene, 40, 41, 65, 86, 88
Polyisoprene
cis (natural rubber), 10, 15, 65, 68, 154
trans (gutta-percha), 10, 88
Polymerization
addition, 21, 22, 23
condensation, 21, 22, 23, 104
general, 19, 24
ionic, 50
template, 140, 141, 144
Polymethylene oxide, 10, 86, 88
Polymethyl methacrylate, 9, 41, 62, 65, 66, 93
Polynucleotides, 21
Polyolefin, 9
Poly m-paramid, 11
Polypeptide, 13, 43, 100, 104, 106, 108, 109, 113, 123
Polyphenylene ethers, 14
Polyphenylene oxide, 65, 87, 88
Polyphenylene sulfide, 14, 87, 88
Poly-L-proline, 40, 41, 105, 108, 110, 117
Polypropylene, 9, 17, 41, 49, 50, 51, 52, 65, 85, 86, 88, 89
Polysaccaharide, 13, 15, 21, 123
Polystyrene, 9, 62, 63, 65, 66
Polysulfide, 10
Polysulfone, 12, 14
Polyurethane, 11, 21, 154
Polyvinyl chloride, 9, 19, 50, 65, 66
Polyvinylidene fluoride, 9
Polytetrafluoroethylene, 9, 14, 19, 48, 58, 87
Proline, 101, 109, 117, 118, 144
Propane, 14
Prosthetic, 154
Proteins,
denatured, 54
fibrous, 54, 113, 114, 116, 120, 121
general, 1, 21, 100, 101, 109, 111, 130, 155
globular, 54, 113, 121, 122, 127
synthesis, 130, 131, 143, 144, 146, 149, 150, 151
Porphyrin ring, 127
Purine bases, 131, 132, 134, 135, 143
Pyrimidine bases, 131, 132, 134, 135, 143

Rayon, 12, 88, 93
Recombinant DNA, 151, 152, 155

Reithel, F.J., 125
Replication, 143, 151
Ribonuclease, 111, 112, 113, 126, 127, 145
Ribosomes, 143, 149, 151
Rice, R.V., 118, 120
RNA (see nucleic acids)
Rogers, G.E., 115
Rubbers, 1, 18, 54, 61, 62, 93

Saran, 18
Schulz, G.E., 156
Seymour, R.B., 156
Shaw, J.T.B., 116
Shrimer, R.H., 156
Shrinkage, 26, 94, 96, 114, 118
Silicone rubber, 10, 88
Silk, 1, 27, 91, 112, 116, 117
Skin, 111, 114
Smith, D.J., 45
Smith, S.G., 116
Smokeless powder, 12
Spandex, 17, 18
Specific heat, 63
Specific volume, 63, 64, 84
Sperulite, 79, 81, 82
Spinneret, 90
Spinning, 89, 90, 91
Staudinger, H., 2, 28
Starch, 12, 13, 15, 16, 17, 21
Steel, 93, 94
Stereo-isomer, 50, 51, 68, 85, 101
Steric influence, 31, 50
β-Structure, 107, 108, 109, 116
Supermolecular structure, 73, 79, 81
Superstructures, 119
Svedberg, T., 28
Syndiotactic, 50, 51, 85

Tadokoro, H., 46
Tanning, 26
Teflon, 14, 48, 88
Tendons, 111, 114, 117
Tensile force, 90
Tensile strength, 89, 93
Tension development, 98, 99
Thermal expansion, 55, 63, 64
Thermal stability, 14, 26
Thermodynamics, 56, 72, 80, 81
  second law, 56
Thermoelastic properties, 55, 61
Thermoplastics, 86
Thymine, 132, 134, 143
Transiton, rubber-glass, 63, 64, 65
Treloar, L.R.G., 35, 39, 58, 62, 156
Tropocollagen, 117, 118, 119

Unit cell, 68, 69, 70, 80
Uracil, 132, 143
Urea-formaldehyde, 25

Veins, 111
Vinylite, 18
Vinyl monomers, 21
Vinyl polymers, 42, 49
Viscose rayons, 12
Voigt-Martin, I.G., 80, 81
Vulcanization, 26

Watson, J.D., 134, 135, 140, 156
Weber, A., 121
Wilkins, M.H.F., 134
Winslow, F.H., 156
Wood, 1, 12
Wool, 1, 26, 27, 91, 93, 94, 112, 114, 116

X-ray diffraction, 68, 73, 114, 122, 123, 134

Young's modulus, 92, 93

Ziegler, K., 50